Undergraduate Texts in Physics

Undergraduate Texts in Physics (UTP) publishes authoritative texts covering topics encountered in a physics undergraduate syllabus. Each title in the series is suitable as an adopted text for undergraduate courses, typically containing practice problems, worked examples, chapter summaries, and suggestions for further reading. UTP titles should provide an exceptionally clear and concise treatment of a subject at undergraduate level, usually based on a successful lecture course. Core and elective subjects are considered for inclusion in UTP.

UTP books will be ideal candidates for course adoption, providing lecturers with a firm basis for development of lecture series, and students with an essential reference for their studies and beyond.

M. A. Curt Koenders

Constructing the Edifice of Mechanics

From Newton to Modernity

Springer

M. A. Curt Koenders
Civil, Maritime and Environmental
Engineering
University of Southampton
Southampton, UK

ISSN 2510-411X   ISSN 2510-4128   (electronic)
Undergraduate Texts in Physics
ISBN 978-3-031-34070-3   ISBN 978-3-031-34071-0 (eBook)
https://doi.org/10.1007/978-3-031-34071-0

This Springer imprint is published by the registered company Springer Nature Switzerland AG
The registered company address is: Gewerbestrasse 11, 6330 Cham, Switzerland

Preface

This book deals with theoretical mechanics. Newton published the *Philosophiæ Naturalis Principia Mathematica* in 1687. In it he sets out the basic principles of physics that are required to understand the motion of the planets, their moons and the comets in the solar system. It includes the gravitational (inverse square) law, the inertial principle and the basic elements of mechanics. Since its publication a large number of refinements and reformulations have been introduced, thereby adding enormous insight in the structure of mechanics, which is commonly known as 'classical mechanics'. All these have in common that by taking a suitable limit, Newton's original principles reappear. Thus, physicists and mathematicians who work on the subject always have a notion that if their theories do not return to Newton's foundations then there is something wrong. Newton himself acknowledged that 'if I have seen further (than others), it is by standing on the shoulders of giants'. One of these giants was undoubtedly Galileo who died in the year Newton was born. So, Newton himself adhered to the 'classical limit'.

Below a table of notable physicists is produced, which is totally incomplete and omits many contributors (apologies are in order). The table gives an impression of the development of physics since Newton. The history of the subject may be delineated in two main sets of ideas that fed the intellectual curiosity of these people. The first pertains to work before 1900, in which their names are associated with major extensions to Newton's work—frequently as a result of improved understanding of differential calculus. The second set is relevant to the break-neck-speed development of theoretical physics after 1900. The advent of relativity theory and quantum mechanics has put a whole new slant on the classical limit. In this book the emphasis is on showing under what conditions that limit returns.

Galileo Galilei	1564–1642
Isaac Newton	1642–1726
Leonhard Euler	1707–1783
Jean-Baptiste (le Rond) d'Alembert	1717–1783

Joseph-Louis Lagrange	1736–1813
Pierre-Simon Laplace	1749–1827
Siméon Poisson	1781–1840
Carl Jacobi	1804–1851
William Hamilton	1805–1865
James Clerk Maxwell	1831–1879
Ludwig Boltzmann	1844–1906
Hendrikus Lorentz	1853–1928
Max Planck	1858–1947
Albert Einstein	1879–1865
Niels Bohr	1885–1962
Werner Schrödinger	1887–1961
Werner Heisenberg	1901–1976
Paul Dirac	1902–1984
John Bell	1928–1990

In some instances the post-1900 physics gives results that would appear to be distinctly un-Newtonian. But, never fear, classical mechanics is always just around the corner. The wonderful thing is that the non-classical results have been systematically vindicated by experiments.

This is a good place to explain what is meant by 'un-Newtonian'. It has been argued that Newton's mechanics (and all the subsequent refinements) do nothing but codify our intuition and in this way it is an expression of 'Western culture'. So, principles such as 'cause has to come before effect' and the inertial principle are deemed to be, in some sense, culture-dependent. Such principles are expressions of common beliefs and may change meaning when travelling to other cultural domains. However, other cultures were also assiduously observing the motion of the planets (Babylonians, Chinese, Incas) and they would also admit that things don't fall upwards. And while the expression of the common belief may have taken various forms, the fascination with the manner in which Nature speaks is universal. What made the Western Way unique and extremely successful was its use of mathematics. It is also worth noting that Newtonian mechanics appeared at the same time as the emergence of the Age of Enlightenment (with René Descartes' *Discourse on the Method* in 1637, with its—at the time, provocative—dictum, 'Cogito, ergo sum'). Newtonian mechanics may have supplied the Enlightenment with nutritious fodder; it is certainly very strongly associated with it.

The role of the church is important. Galileo was investigated by the Inquisition for supporting the heliocentric view of the solar system. By the time Newton wrote the *Principia* the Reformation had run its course and the grip of the church on intellectual life was severely diminished, at least in the capitals and centres of learning in Western Europe. Lagrange was born Italian, worked in Berlin (on the recommendation of the Swiss mathematician Euler and the French scientist d'Alembert) and later in Paris. He wrote the standard work on analytical mechanics *Mécanique analytique*, which appeared in 1788, just over a century after the *Principia*. It remained influential until well into the nineteenth century. By then mechanics had attracted the attention of the military who valued its insights in ballistic problems. Lagrangian mechanics is still the cornerstone of the subject and the canonical formalism is fundamental for every branch of it.

The further elucidation of its mathematical structure is due to Hamilton. Hamilton was eight when Lagrange died, so the two never met (anyway, Hamilton lived in Dublin and Lagrange in Paris); nevertheless Lagrangian mechanics and Hamilton's principle of least-action are very much interlinked, together with the insights by Poisson (Paris) and Jacobi (Berlin). Historians point out that Hamilton had read most of the *Principia* by the time he was 16. This shows that some 130 years after its publication it was still a 'must-read' for budding natural philosophers. Lagrange, d'Alembert and Laplace were sometime collaborators.

Now, illustrating how far analytical mechanics had strayed from the views of the church, Laplace wrote (1814) his *Essai philosophique sur les probabilités*:

> We may regard the present state of the universe as the effect of its past and the cause of its future. An intellect which at a certain moment would know all forces that set nature in motion, and all positions of all items of which nature is composed, if this intellect were also vast enough to submit these data to analysis, it would embrace in a single formula the movements of the greatest bodies of the universe and those of the tiniest atom; for such an intellect nothing would be uncertain and the future just like the past would be present before its eyes.

This idea, that the Universe evolves completely deterministically, running as clockwork, must be regarded as the pinnacle of a sense of intellectual power (some would say, hubris) that had grown in the eighteenth and nineteenth centuries. It was a hyper-rational opinion. In modern twenty-first-century eyes Laplace's statement must look a trifle arrogant. However, at the time quantum mechanics had not entered the mind of Man yet. Was the statement a deliberate repudiation of religious thinking? Maybe; it was certainly not very diplomatic in a country where 99% still went to Mass every Sunday and the vast majority of the population believed that God—and not the laws of physics—ruled every detail of the Universe.

What the rationalists *did* believe was that every addition, each new insight, every refinement and extension that was made to mechanics should be consonant with Newtonian fundamentals. This belief in the classical limit has endured. From the beginning of the nineteenth century until the present day, workers in theoretical physics are steeped in classical mechanics. Its methods, conservation laws and the quantities in which physical entities are expressed—momentum, energy, etc.—have all been pioneered in classical mechanics. Their use has been endlessly validated by numerous, increasingly accurate laboratory experiments and celestial observations. It is exactly these experiments that have thrown a spanner in the works of classical mechanics. It is difficult to overstate the intellectual effort that was required to come to terms with the 'new' facts that emerged at the end of the nineteenth century and the beginning of the twentieth. The comfortable—intuitively correct—world of classical mechanics was challenged by ideas that would appear to be completely counter-intuitive (Russell McCormmach's *Night Thoughts of a Classical Physicist* gives some insight in the struggles that went on physics departments). So, men like Einstein,

Bohr, Heisenberg and Schrödinger, while exploring ways in which experiments could be accommodated by suitable theory, used Newtonian, Lagrangian and Hamiltonian concepts as a solid foundation. At the same time they had to confront the fact that *in certain instances classical mechanics does not work*. It transpired that, rather than the 'new' physics being an extension of classical physics, a much greater—more encompassing—theory had to be set up, in which classical physics is an island that is valid in certain limits. Two examples are relativity theory and quantum theory, both initially conceived as separate theories. It fell to Dirac to perceive a mathematical structure that unified these two into one overarching framework. (Dirac's work is largely incomprehensible for the non-expert, which may be the reason that he never got any popular recognition; in fact he was rather removed from popular culture and you would not expect him to appear on, for example, *Desert Island Discs*. The giant leaps he took prompted Ehrenfest to state that his physics was 'inhuman'. Graham Farmelo's biography of Dirac *The Strangest Man* is a must-read for everyone with an interest in how progress in theoretical physics is made.)

A troublesome issue for quantum mechanics, its measurement and its interpretation arose as early as 1935 in a paper by Einstein, Podolsky and Rosen. At its heart was a classical question: if two particles have interacted in the past can a measurement of a property on one of them instantaneously define the outcome of a measurement on the other one, no matter how far apart they are at the time of measurement. An obvious paradox, this led to rather muscular discussions. In 1965 John Bell published a paper (in a rather non-mainstream journal) in which he described the exact difference between experiments on classical objects and quantum objects and suggested a putative test to establish the difference. A first experiment to test the theory was done in 1972 by Freedman and Clauser, subsequently refined by others, see also Section (8.9.6.1). The outcome vindicated the quantum mechanics view. While this leaves many more questions about the exact nature and the correct interpretation of quantum theory, it is noted that Einstein, Podolsky and Rosen's paper was based on a question from the realm of classical physics. The answer, nearly 40 years later, will have been unsatisfactory to some. It shows how important the classical limit still is for the progress of physics. However, there is a subtle shift, because the basis of Bell's paper did not appeal to classical *mechanics*, but to classical *intuition*.

That is where the subject stands. No doubt much more ink will be spilled on the implications and—indeed—applications for quantum and classical physics and many more ideas will be put forward. It is a wonderful, entertaining and stimulating subject.

The motivation for writing this book was foremost as a service to a group of students who studied mathematics and want to do further study, at master's or doctoral level, in physics. Traditional books on mechanics tend to incorporate an introductory text on mathematics. Here the mathematics is much less in the foreground, but rather readers are introduced to concepts in physics. The classical limit was chosen as the leitmotiv, as it would be the key to many of the questions that researchers in the field have when they manufacture a new piece of theory: 'How does this new theory converge to the old theory, which has been extensively tested by experiment?' A lot of physics, especially classical mechanics, is based on intuitive notions and in producing a new theory, or by putting forward an extension of an older one, the researcher needs to show exactly where and why it is different.

For the category of readers who have no background in mathematics (engineers, economists or the more mathematically based life science practitioners) and who want to enjoy this study, there is a mathematical appendix, which they may find useful.

The approach to the subject matter in the book is inspired by a remark overheard in Canterbury in a museum. A French family with two children visited and looked around. Then the father said: 'C'est comme chez nous', or 'just like home'. In this book a journey through the Land of Mechanics is undertaken. All the while the travellers look out through the windows of the train and compare what they see with what they know from 'back-home'. The classical limit in mechanics is somewhat like that; new terrain is explored and new impressions acquired, but the explorer instinctively compares the 'new' with the 'old'. That is the essence of what makes the 'new' understandable and the subject exciting.

Canterbury, UK   M. A. Curt Koenders
April 2023

Contents

Chapter 1
Newtonian Mechanics

Abstract Newton's laws of motion and Newton's gravitation law are introduced; the wider context with an eye to modern physics of these is discussed. The notion of the classical limit is presented. Conservation laws that can be derived from the laws of motion are presented. Then examples of one-dimensional motion are treated. The treatment is restricted to point masses. The stability of motion is illustrated using the Lyapunov function and the concept of damping is analysed for the example of the harmonic oscillator. An introduction to the subject of colliding bodies is given and finally the mechanics of disintegrating bodies is studied.

1.1   Initial Considerations

Newtonian mechanics deals with the motion of everyday objects in the everyday experience world. Since its inception—Newton's *Principia* was first published in 1687 (Newton 1687, 1995)—there have been very substantial additions, reformulations and extensions. It is therefore reasonable to indicate a validity range for the theory. Informed by twenty-first-century physics that is quite an easy exercise. First, the objects must move appreciably slower than the speed of light. For speeds that approach the speed of light, relativity theory must be invoked and the classical concepts of space and time that underlie Newtonian, or classical, mechanics need to be re-examined. Second, the objects must not be too small, well-above atomic-size. For smaller objects twentieth-century quantum mechanics is the appropriate theory to describe their motion as a function of space and time. Third, space-time must not be too curved, that is, Newton's classical theory needs substantial amendment in the vicinity of very heavy, dense objects. The theory of general relativity is the relevant theory in this case. For example, the motion of the planet closest to the Sun, Mercury, is subject to measurable, though small, non-Newtonian corrections that for a long time remained unexplained, until Einstein's understanding of gravity and its impact on the very structure of space-time, provided a satisfactory description.

While it would be advantageous to be able to describe the motion of physical objects on all scales and under all circumstances, the validity range of Newtonian mechanics is enormous. This is the principal reason why classical mechanics is still worth studying. The other reason is that if an extension to the theory is put forward, it must be possible to indicate what the 'classical limit' is. So, for example, in Einstein's theory of special relativity the equations of Newtonian mechanics can be shown to be valid in the limit of small velocities. Similarly, in quantum mechanics there is the so-called 'correspondence principle', which shows when quantum theory converges to classical mechanics.

'Everyday objects' have a finite size. However, the discussion on classical mechanics is greatly assisted by the introduction of the concept of a *point mass*. As the term implies, a body of finite extent is conceived of as if the whole mass is represented by a point. At a later stage, it will be shown that the most convenient location for this point is the centre of gravity of the body. The point is identified by coordinates. For the moment Cartesian coordinates will be used, which requires the choice of an origin and the notion of direction. The discussion of an appropriate choice of the coordinate frame in classical mechanics will be postponed until the basic concepts of the theory have been elucidated. Meanwhile, a frame is chosen that is fixed with respect to the distant stars. That may appear somewhat impractical; why not fix it to the Sun or even to a suitable place on Earth? Note then, that classical mechanics is not a strictly *mathematical* theory. It makes statements about *physics* and as such puts forward certain notions about the Cosmos. The very idea of a 'cosmological model', as it is understood nowadays (an ever-expanding multi-dimensional, finite object that started as a 'big bang'), was completely outside the scope of the fathers of classical theory: Galileo, Newton, or Leibnitz. They required something that was—or appeared to be—'fixed' and the distant stars fitted the bill.

Once a coordinate frame is chosen, the mass point can be identified by a vector **r**, see Sect. A.1 and Boas (1983), Kibble (1985). A variety of notations is commonly used: a vector has components (r_1, r_2, r_3), or alternatively (x, y, z), or even $\mathbf{x} = (x_1, x_2, x_3)$; these may be functions of the time t. In principle this is all that is needed to describe the motion of the point mass. Its *velocity* **v** is found by differentiation with respect to time: $\mathbf{v} = d\mathbf{r}/dt$. A notation often employed to denote differentiation with respect to time is the 'flux dot', which is a dot over the quantity to be differentiated, so $\mathbf{v} = \dot{\mathbf{r}}$. A second differentiation gives the rate of change of the velocity, the *acceleration* of the object: $\mathbf{a} = \dot{\mathbf{v}} = \ddot{\mathbf{r}}$.

These definitions are sufficient to formulate the basic concepts of classical mechanics.

1.2 Newton's Laws

The basic concepts of classical mechanics are stated in three, rather simple, principles. They are called *Newton's laws*. The first one codifies the so-called *inertial principle*. Originally this principle is due to Galileo.

A body persists in its motion unless a force is exerted on it. The word 'persists' here really means that when a body has a velocity and no forces are exerted on it, then that body will keep going with the same velocity *in a straight line*. The line is straight

Fig. 1.1 Inertial frames

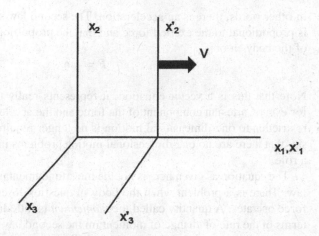

with respect to the distant (fixed) stars. It would be wrong to think of the distant stars keeping track of every little body of the universe and keeping its motion straight. It is equally wrong to think of the body in question keeping a constant watch on where the fixed stars are and triangulating its motion accordingly. The inertial principle is really a property of the universe. Everybody is subject to it, but its explanation is far from clear.

A coordinate frame that is referenced to the location of the fixed stars is called an *inertial frame*. The relation between time and space measurements in two inertial frames, in which one moves with a constant velocity $V = (V, 0, 0)$ with respect to the other is easily established. In Fig. 1.1 the idea is illustrated. Observers located in the origins of both frames carry clocks that are synchronised such that at time $t' = t = 0$ the origins coincide. For simplicity the x_1 coordinates are chosen to align, so the other coordinates remain invariant.

Now, consider an event that takes place at location $\mathbf{x}$ for the unprimed observer and at location $\mathbf{x}'$ for the primed observer. The question is: what is the relationship between the locations and observed time of the event for both frames? The answer is simple: $x_1' = x_1 - Vt$, $x_2' = x_2$, $x_3' = x_3$ and $t' = t$. Incidentally, $v_1' = v_1 - V$.

This transformation is called the *Galileo transformation*. It is the classical transformation and embedded in it is the idea of *universal time*, that is all clocks—no matter what inertial frame they are in—(and assuming that they are all manufactured to the same standard of precision) run synchronously. Not until the beginning of the twentieth century, with the advent of Einstein's relativity theory, was this hypothesis profoundly questioned and, indeed, profoundly altered. Nonetheless, there is the classical limit that says: while the magnitudes of velocities are small compared to the speed of light, relativity theory should converge to classical theory and for time measurement universal time is adequate.

Newton's *second law* is famous and deals with the situation where the first law is violated. A body will *not* persist in its motion, because a force is exerted on it. If the body does not persist in its motion then that implies that the velocity must change;

in other words, there is an acceleration. The second law states that the acceleration is proportional to the exerted force and that the proportionality constant is the *mass* of the body, m or

$$\mathbf{F} = m\mathbf{a} \tag{1.1}$$

Note that this is a vector equation; it represents really three equations and holds for every Cartesian component of the force and the acceleration. Crucially then, the restriction to one-dimensional motion is no longer required. This does not mean to say that there are no one-dimensional motion problems in which accelerations play a role.

The equation as given here is not the most fundamental form of Newton's second law. There is a problem when the body in question loses or gains mass while the force operates. A quantity called the *momentum* is introduced, which is $\mathbf{p} = m\mathbf{v}$. In terms of the rate of change of momentum the second law reads

$$\mathbf{F} = \frac{d\mathbf{p}}{dt} \tag{1.2}$$

When the mass is constant this obviously reverts back to the form given above.

A number of examples of the second law that are important in other branches of physics will be given in what follows.

Newton's *third law* deals with interactions between point masses. One point mass may exert a force on another point mass. Newton's third law states simply that the force from the first onto the second is equal and opposite to the one exerted by the second onto the first. This law is sometimes encapsulated in the pithy phrase 'action equals minus reaction'. However, the words 'action' and 'reaction' have other meanings in mechanics, so this may not be a very helpful phrase.

These are Newton's laws. They are the first basis for classical mechanics and permit the calculation of many problems with accurate experimental verification. The latter would include the measurement of masses and forces, in addition to displacements and time. Now, the measurement of mass and force could easily lead to a tautology. One could either measure force by subjecting a mass to an acceleration, or measure a quantity of mass to a known force. Either mass or force has to be defined independently for the measurement to make sense. So, an extra element is required. For example one could define 'force' by relating it to the extension of a standardised spring. For accurate measurement that is obviously difficult, as much would depend on the reproducible manufacture of the spring. However, it demonstrates that in order to employ Newton's laws of motion the force has to be given by phenomenological means. The problem is solved to some extent by Newton's law of gravitation. This law states that two masses attract one another, depending on the distance between them. More precisely, for two masses m_1 and m_2 the force of attraction is inversely proportional to the square of the distance r between them:

$$F_{12} = G \frac{m_1 m_2}{r^2} \tag{1.3}$$

While this law defines the extra relationship between mass and force and so—in principle—fixes the tautology, various other problems appear. The first is the difficulty of knowing that the m in the second law is the same as the m in the gravity law. They need not be. The mass in the second law is called the *inertial mass*; the masses in the gravitation law are called *gravitational masses*. The difference between the two is open to experimental verification by creating a laboratory set-up in which both play a role. In particular, the Hungarian physicist Eötvös performed a sensitive test and could not, within his measurement accuracy, find any difference between the two. Einstein's theory of general relativity uses this finding as its very basis and, conversely, the experimental tests to which the predictions of this theory have been subjected confirm the equivalence of inertial and gravitational mass. So, while *empirically* they are identical, *ontologically* they are different concepts.

The second issue that arises from the law of gravity is the constant G, the *gravitational constant*. Its measurement is rather fraught, mainly because the interaction is so weak and different experiments give different answers after the fifth decimal place, see also Sect. 4.1. The current accepted value is $G = 6.67408 \times 10^{-11} \text{ m}^3 \text{ kg}^{-1} \text{ s}^{-2}$. The weakness of the gravitational interaction is illustrated by comparing it to the electrical interaction at the same distance for two elementary particles. For two protons, for example, the ratio of the electrical force to the gravitational force is some 10^{36}. In many situations for macroscopic bodies, however, the electrical force plays no significant role, because the charge in matter is perfectly balanced—there is as much positive as negative charge—so it appears neutral. It is then reasonable to calculate the force a body on the surface of the Earth experiences. The mass of the Earth is $5.9724 \times 10^{24} \text{ kg}$, the volumetric mean radius is 6371 km, which, using the gravitational law, yields a force per unit mass of 9.82 Nkg^{-1}. In other words, using the equivalence of inertial and gravitational mass, the acceleration on the surface of the Earth is 9.82 ms^{-2}. This value is called g. For various other celestial bodies in the solar system that have different masses and radii the value of g is as follows, for the Moon it is 1.6 ms^{-2}, on Mars it is 3.8 ms^{-2} and on Jupiter 26 ms^{-2}. Given the value of g, the unit of mass can be calibrated by comparing it to an internationally agreed standard mass. In Sèvre, a suburb of Paris, a platinum-iridium block is kept in a glass cage for this purpose. It is the Société Internationale (SI) definition of the kilogram. Copies can be ordered by other laboratories. In 2019 the definition was changed and tied to subatomic quantities, which involves fixing fundamental physical constants. The latter are then *defined*, rather than measured by comparison to material standards.

1.3 Conservation Laws

Newton's second law can be employed to calculate the orbit of an object, given the force environment and initial conditions for the position and the velocity. The second law leads to a differential equation. It is a reasonable question to ask whether there are conditions under which the equation has an integral. These would then lead to conservation laws.

Conserved quantities are extremely important in mechanics. Processes are frequently described in terms of the progress, transfer or flow of a conserved quantity. Even in everyday, non-scientific discourse it is encountered. For example, a weatherman may say that 'there is a lot of energy in the atmosphere that is released in thunderstorms'. Here the transfer of 'energy' is used to describe a weather event. There is no doubt that energy is a valid concept in this case, but there is absolutely no way of measuring the energy. Some of it may be in the flow of airmass, some may be present in the heat of, for example, a nearby body of water. However, it is convenient to describe the process in terms of a simple property and the weatherman's audience would find it a plausible metaphor to help them understand on some level what the progress of the weather will be. Similarly, a commentator on financial matters may find it helpful to represent a quantity of money as a conserved quantity that flows from a certain source to some other destination.

> In fact, an analogue computer system was set up by an economist by the name of *Bill Phillips* who developed the 'Phillips machine', which is a device in which water is pumped round to various reservoirs to illustrate effects in the national economy: so much goes to companies, so much to individuals, so much in tax, etc. The moving water is deemed to represent the flow of money and its sum total is supposed to be a conserved quantity (no bank notes were supposed to be burnt). It is of interest to note that the leaks in the system were assumed to represent the black economy, with water ending up on the floor of the basement of the London School of Economics.

In physics there is no place for didactic metaphors. The conserved quantities correspond to exactly measurable features of a system and their conservation as a function of time must be rigorously true. It follows from the equations that rule the processes. This is what makes them such a valuable tool in mechanics.

Some conserved quantities in physics are not derived from mechanical laws. For example, in electrodynamics *charge conservation* is a valid concept; essentially charge can be neither destroyed nor created (though it can be redistributed in space as a function of time) and this idea is codified in the *equation of continuity*. Similarly, in classical mechanics the conservation of mass holds. Even though there may be chemical changes that occur in a process, the total number of protons and neutrons that participate in a process does not change. (Nuclear processes do not come under classical physics.)

Below the interest is first of all focussed on conservation laws that can be derived from Newton's laws, as well as the conditions under which they are valid.

1.3.1 Conservation of Momentum

The rather simple case of no external forces working on a body leads to the conclusion that

$$\frac{d\mathbf{p}}{dt} = 0 \qquad (1.4)$$

So, when there are no external forces momentum is conserved.

An example would be a system of two bodies with masses m_1 and m_2, located at $\mathbf{r}_1$ and $\mathbf{r}_2$. The centre of gravity of this system is located at

$$\mathbf{R} = \frac{m_1 \mathbf{r}_1 + m_2 \mathbf{r}_2}{m_1 + m_2} \tag{1.5}$$

The two equations are

$$m_1 \mathbf{a}_1 = \mathbf{F}_1 \quad m_2 \mathbf{a}_2 = \mathbf{F}_2 \tag{1.6}$$

Now, note that if these two equations are added together and the assumption is enforced that there are no external forces on the system, so that the only forces at play are the ones from body 1 to body 2 and vice versa, then by Newtons third law the two forces cancel out, so

$$m_1 \mathbf{a}_1 + m_2 \mathbf{a}_2 = 0 \tag{1.7}$$

The centre of gravity is differentiated twice with respect to t and the result multiplied with $m_1 + m_2$ and it follows that

$$\ddot{\mathbf{R}} = 0 \quad or \quad \dot{\mathbf{R}} = constant \tag{1.8}$$

The analysis can easily be extended to a system of many bodies and the result is the same: *for a system on which no external forces are applied the centre of gravity persists in its motion.* This conservation law holds in vector form, so it represents really three conservation equations.

The first derivative with respect to time of the centre of gravity is easily recognised as the total momentum of the system divided by the total mass.

The concept of a point mass can now be revisited. A solid body of finite extent can be regarded as a collection of point masses. These all interact, but the motion of the whole body can be viewed as that of one mass centred at the centre of gravity. Therefore, the concept of a point mass may be extended to a finite-sized body.

1.3.2 Conservation of Energy

A scalar equation can be created by taking the inner product (see Sect. A.1, Boas 1983; Kibble 1985) of the second law with the velocity:

$$\mathbf{F} \cdot \dot{\mathbf{r}} = m \ddot{\mathbf{r}} \cdot \dot{\mathbf{r}} \tag{1.9}$$

The right-hand side is easily recognised as $d(\frac{1}{2}mv^2)/dt$; the quantity $\frac{1}{2}mv^2$ is called T, which is known as the *kinetic energy*.

In order to arrive at a conservation law the left-hand side must also be the time derivative of a quantity. This can be achieved if the force can be written as a spatial derivative of another quantity V, in other words if $\mathbf{F} = -\text{grad } V$ (the minus sign is convention). The *potential energy* V is a scalar. Now, the left-hand side may be recast in the form

$$\mathbf{F}.\dot{\mathbf{r}} = -\frac{dV}{dt} \tag{1.10}$$

Energy conservation follows from

$$\frac{d(T+V)}{dt} = 0 \ \ \text{or} \ \ T+V = constant \tag{1.11}$$

For this conservation law to be valid, the force must be able to be written as a gradient of the potential energy. This means that the force depends on the position only. The physical implications of such an assumption are investigated by considering very slow, so-called 'quasi-static', displacements. A slow, incremental movement of the body is denoted by $\delta\mathbf{r}$; a force field $\mathbf{F}$ is in operation. An increment of *work* is defined as

$$\delta W = \mathbf{F}.\delta\mathbf{r} \tag{1.12}$$

The orbit of the body, carried out in such a way that it ends up in the same place, represents an amount of work

$$W = \oint_s (\mathbf{F}.\delta\mathbf{r}) \tag{1.13}$$

where the integral is over a closed contour s. The integral can be recast to one over a surface S, using Stokes' theorem (see Sect. A.3)

$$W = \iint_S (\nabla \times \mathbf{F}).\mathbf{n}dS \tag{1.14}$$

Now, if the force can be expressed as the gradient of V the result is zero, as $\nabla \times \nabla V = 0$. Therefore, *if the force can be expressed as a potential energy gradient, no work is done when a body has moved through a closed orbit*. When that is the case the force is said to be a *conservative force*.

The conservation of energy is closely associated with *thermodynamics*. Mechanical work represents one form of energy. However, heat (including radiation) is also a form of energy. This is codified in the *first law of thermodynamics* for a closed system, which requires the introduction of the concept of *internal energy*, U, which represents the total energy of the 'system' and is closely related to its temperature. A change in this quantity, δU, can be effected by either an increment of mechanical work δW or an increment in heat δQ. The rules that govern the transfer of mechanical energy to heat are the province of thermodynamics, which will be returned to in the study of *statistical mechanics*, a branch of physics which is concerned with the study of systems that consist of many bodies.

1.3.3 Angular Momentum

Angular momentum is defined as $\mathbf{L} = \mathbf{r} \times \mathbf{p}$. In certain circumstances the angular momentum may be a conserved quantity. To find out what those circumstances are, evaluate its time derivative

$$\frac{d\mathbf{L}}{dt} = \dot{\mathbf{r}} \times \mathbf{p} + \mathbf{r} \times \dot{\mathbf{p}} \tag{1.15}$$

Now, $\mathbf{p}$ and $\dot{\mathbf{r}}$ will for a single body be in the same direction; therefore the outer product will vanish. Applying Newton's second law then leads to

$$\frac{d\mathbf{L}}{dt} = \mathbf{r} \times \mathbf{F} \tag{1.16}$$

It follows that if the force is always directed in the direction of the position vector then the outer product vanishes and angular momentum is conserved. In these cases the force emanates from a single point; it is a so-called *central force*. Such a force is very important in physics. It is easily seen, for example, that the motion of the planets around the Sun is ruled by a central force due to the gravitation law.

The quantity $\mathbf{r} \times \mathbf{F}$ is known as the 'force moment'. For bodies that stand still the rules of *statics* apply: sum of forces and sum of force moments must vanish.

The conservation of momentum and angular momentum will be studied in more detail later as a reformulation of classical mechanics is undertaken.

1.4 Examples of Motion in One Dimension

In one-dimensional motion Newton's equations yield simple differential equations. A few examples are treated here that give insight in simple mechanical processes and show the constitutive concepts that have to be introduced to describe them.

1.4.1 Object in a Constant Gravity Field

In a constant gravity field the force on an object with mass m is, as has been seen above equals to gm. This is a good approximation for objects that do not stray too far from the surface of the Earth, for example. Denoting the position of the object by z, the second law takes the form

$$m\ddot{z} = -gm \tag{1.17}$$

The mass of the object drops out of this equation, as inertial and gravitational mass are notionally indistinguishable. The solution of the equation is then simply

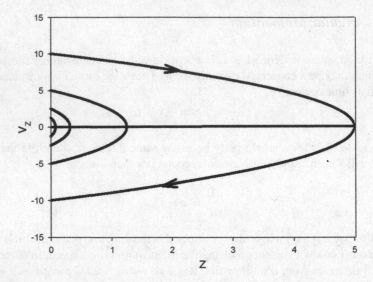

Fig. 1.2 Motion of a bouncing body depicted in phase space

$$z(t) = z_0 + \dot{z}_0 t - \frac{1}{2}gt^2 \qquad (1.18)$$

A special case is the one in which the object is shot from the ground with an initial velocity. Then $z_0 = 0$ and only $\dot{z}_0$ needs to be specified. The object flies upwards until it reaches its highest point (at $\frac{1}{2}\dot{z}^2/g$); then it reverses velocity and falls back. The question is then: what happens when the object hits the ground? The answer is: it bounces. The velocity is suddenly reversed to a fraction $-c_r$ of the impact velocity. The *restitution coefficient* c_r is a material constant that depends uniquely on the materials of which the object and the 'ground' are made. The easiest way to visualise the process is in so-called *phase space*; here the velocity is plotted as a function of the position. The time coordinate is thus eliminated. Figure 1.2 gives an idea of how this works.

The object in the figure starts its orbit in phase space at ground level with a velocity of $10 m/s$; then it follows the arrows till the z-coordinate reaches 0 again. The velocity reverses and the body starts another cycle, up and down, until it reaches the ground again and reverses velocity once more. This happens a number of times, each subsequent cycle being smaller and smaller. The energy function for each cycle is $\frac{1}{2}m\dot{z}^2 + mgz$ and its value is therefore given by $\frac{1}{2}m\dot{z}_0^2$. The orbits in phase space are given by $\dot{z} = \pm\sqrt{\frac{1}{2}m(\dot{z}_0^2 - \dot{z}^2) - mgz}$.

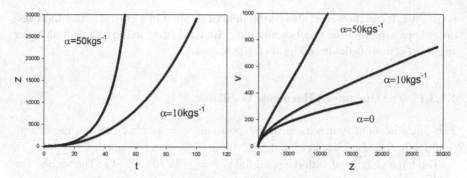

Fig. 1.3 Motion of a rocket that loses mass depicted as the displacement as a function of time and in phase space

1.4.2 A Rocket That Expels Mass

This example is included to demonstrate how the second law is phrased when the mass is not constant. The momentum form of the second law is then required. The problem is phrased as follows. A rocket with initial mass m_0 is launched vertically in a gravity field with acceleration g. The thrust of the rocket engine is assumed constant and exerts a force F, while the rocket engine burns fuel and expels exhaust gas (and thus loses mass) at a constant rate α. The mass of the system is therefore described as $m = m_0 - \alpha t$.

The second law reads

$$\frac{dp}{dt} = \frac{d\left[(m_0 - \alpha t)\dot{z}\right]}{dt} = -(m_0 - \alpha t)g + F \tag{1.19}$$

This equation has a solution under the initial conditions $z_0 = 0$, $\dot{z}_0 = 0$, as follows

$$z(t) = -\frac{1}{4}gt^2 - \frac{t}{2\alpha}(2F - gm_0) - \frac{m_0}{2\alpha^2}(2F - gm_0)\ln\left(\frac{m_0 - \alpha t}{m_0}\right) \tag{1.20}$$

The force of the thrust F obviously needs to be larger than the weight of the rocket at launch $m_0 g$ for the thing to go up. However, once it goes it keeps accelerating because the mass decreases. This continues till the fuel is spent. By way of example use $g = 10\,\mathrm{ms^{-2}}$, $m_0 = 3000\,\mathrm{kg}$ and $F = 40000$ N. Depending on the value of α then the rocket accelerates as a function of time as illustrated in Fig. 1.3.

1.4.3 The Harmonic Oscillator

The harmonic oscillator is very important in physics. Its essence is simply an object in a quadratic potential field. This field can be thought of as an 'ideal' spring with spring constant k. It can, however, also be the field of an atom in a crystal lattice in lowest order approximation for small oscillations. Similarly, the motion of a pendulum for small deflections is described by such a potential.

1.4.3.1 The Undamped Harmonic Oscillator

The potential field is quadratic in the position: $V = \frac{1}{2}kx^2$; therefore, the energy function is $E = \frac{1}{2}m\dot{x}^2 + \frac{1}{2}kx^2$ and the orbit in phase space is an ellipse.

The force associated with the potential is $F = -\partial V/\partial x = -kx$. The second law reads

$$m\ddot{x} = -kx \tag{1.21}$$

This has a solution with two initial conditions x_0 and $\dot{x}_0$

$$x(t) = x_0\cos(\omega t) + \dot{x}_0\sin(\omega t) \tag{1.22}$$

where the *circular frequency* $\omega = \sqrt{k/m}$. The circular frequency is related to the *frequency* f as $\omega = 2\pi f$ and the *period* of oscillation is $T = 1/f$. The orbit of the oscillator in phase space is depicted in Fig. 1.4.

It is of interest to find out what the *probability* is of finding the object in a certain place. The probability $P(x)dx$ of finding it in an infinitesimal region dx at x is proportional to the time it spends there and so

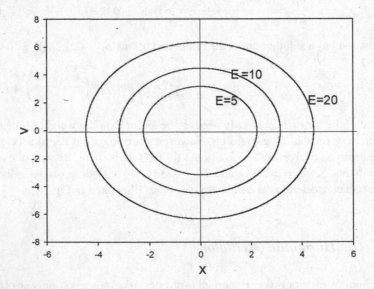

Fig. 1.4 Motion of a harmonic oscillator in phase space

$$P(x)dx \propto dt \qquad (1.23)$$

In other words

$$P(x) \propto \frac{dt}{dx} = \frac{1}{v} \qquad (1.24)$$

The velocity needs to be expressed as a function of x. Suppose that $x_0 = 0$, then $x(t) = A \sin \omega t$ and $v(x) = A\omega\sqrt{1 - x^2/A^2}$. Therefore,

$$P(x) \propto \frac{1}{A\omega\sqrt{1 - x^2/A^2}} \qquad (1.25)$$

The proportionality constant is determined by requiring that the probability of finding the oscillator between the two extremes of the position is unity

$$\int_{-A}^{A} P(x)dx = 1 \qquad (1.26)$$

And it follows that

$$P(x) = \frac{1}{\pi\sqrt{A^2 - x^2}} \qquad (1.27)$$

In Fig. 1.5 it is seen that the probability goes to infinity at the turning points, when the velocity is zero and obviously most time of the oscillator is spent in these points.

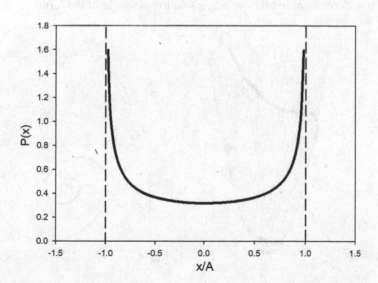

Fig. 1.5 Probability function $P(x)$

1.4.3.2   The Damped Harmonic Oscillator

A refinement is implemented by introducing *damping*. An extra element is added to
the force that is velocity dependent: $F = -kx - \gamma\dot{x}$. Now there is no energy integral,
but it is still useful to see what happens if the second law is multiplied by $\dot{x}$.

$$m\dot{x}\ddot{x} + kx\dot{x} = -\gamma(\dot{x})^2 \;\rightarrow\; \frac{d}{dt}(\frac{1}{2}m\dot{x}^2 + \frac{1}{2}kx^2) = -\gamma\dot{x}^2 \qquad (1.28)$$

The right-hand side of this equation is always negative; therefore, the energy $E =
\frac{1}{2}m\dot{x}^2 + \frac{1}{2}kx^2$ always goes down until the phase point reaches the origin. Depicting
the energy as a function of position and velocity gives a parabolic bowl and the
phase point spirals towards the origin. Such a bowl is called a *function of Lyapunov*.
It indicates that there is an attraction point where, whatever the starting point, the
system ends up. So it says something about the stability of a system.

The concept of the Lyapunov function is illustrated in Fig. 1.6. The projection
onto the phase plane is also shown.

A further extension to the theory of the harmonic oscillator is an external agitating
force. For the moment its exact form is left unspecified. Instead, the response of the
system to a delta function is considered, see Sect. A.6. Thus, try to solve

$$m\ddot{G} + \gamma\dot{G} + kG = \delta(t - t') \qquad (1.29)$$

This describes the agitation by a pulse at the time $t = t'$. A solution is obtained by
Fourier transformation; in other words, a solution is sought of the form

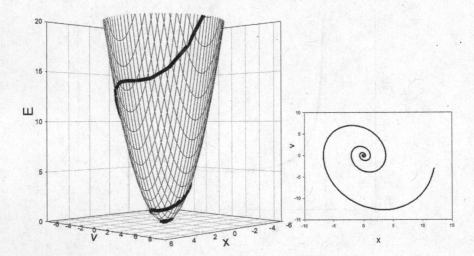

Fig. 1.6 Parabolic bowl with the orbit of the phase point for a damped harmonic oscillator and
projection on the phase space

$$G(t, t') = \frac{1}{2\pi} \int_{-\infty}^{\infty} \hat{G} e^{i\omega t} dt \tag{1.30}$$

The delta function is written as

$$\delta(t - t') = \frac{1}{2\pi} \int_{-\infty}^{\infty} e^{i\omega(t-t')} dt \tag{1.31}$$

Substituting in the differential equation gives

$$(-m\omega^2 + i\gamma\omega + k)\hat{G} = e^{-i\omega t'} \tag{1.32}$$

Therefore,

$$\hat{G} = \frac{e^{-i\omega t'}}{-m\omega^2 + i\gamma\omega + k} \tag{1.33}$$

So that

$$G(t - t') = \frac{1}{2\pi} \int_{-\infty}^{\infty} \frac{e^{i\omega(t-t')}}{-m\omega^2 + i\gamma\omega + k} d\omega \tag{1.34}$$

This solution is called a *Green's function*. Its significance is as follows. If the agitation is not a delta function, but an arbitrary function of time $f(t)$, then using the property that

$$f(t) = \int_{-\infty}^{\infty} f(t')\delta(t - t')dt', \tag{1.35}$$

the solution for the displacement is

$$x(t) = \int_{-\infty}^{\infty} f(t')G(t - t')dt' \tag{1.36}$$

For example, if the oscillator is agitated by an external periodic force with amplitude A and circular frequency ω_0, so $f(t') = A\cos(\omega_0 t')$. By doing the integral it is easily shown that with $\bar{\gamma} = \gamma/m$ and $\omega^2 = k/m$

$$x(t) = x_0 \cos(\omega_0 t - \phi) = \frac{A\cos(\omega_0 t - \phi)}{m\sqrt{(\omega^2 - \omega_0^2)^2 + \omega_0^2\bar{\gamma}^2}} \tag{1.37}$$

where

$$\tan\phi = \frac{\bar{\gamma}\omega_0}{\omega^2 - \omega_0^2} \tag{1.38}$$

This shows that for an agitation of frequency equal to the natural frequency the amplitude of the displacement x_0 is maximal, while it decays for frequencies away from ω. When the damping is zero the response at the natural frequency is infinite.

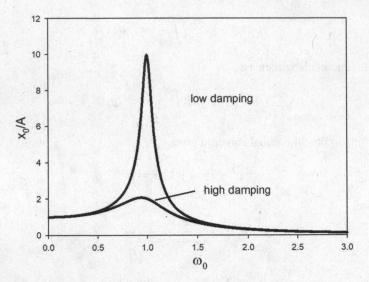

Fig. 1.7 Frequency-dependent response of a damped harmonic oscillator

The magnitude of the response also depends on the damping, as illustrated in the figure of the *frequency characteristic*, the ratio between the agitation amplitude and the response amplitude. Figure 1.7 gives an example. These concepts return in many branches of physics, for example, in analog electronics.

1.4.4   Friction

When two solid surfaces are pressed together with a force F_n and then a tangential force F_t is applied and slowly increased from zero, it is found that initially the two surfaces offer great resistance to the tangential force. When the tangential force reaches a critical value, however, very suddenly the resistance ceases. Coulomb has provided a criterion for the critical value, which is often expressed as a tangent of the so-called friction angle ϕ_μ. Coulomb's criterion is

$$\frac{F_t}{F_n} = \tan \phi_\mu \qquad (1.39)$$

The frictional phenomenon, which is very important in engineering (for example, slope stability problems), rather defies an easy description in terms of displacements or velocities. A description that is usable is an incremental force-displacement description, see Koenders (2020) for further reference.

1.4.5 Collisions and Disintegrating Objects

1.4.5.1 Collisions

The mechanics of two colliding point masses is easily treated using the conservation laws of momentum, angular momentum and energy and the concept of the coefficient of restitution. Let two masses m_1 and m_2 come together with velocities $\mathbf{v}_1$ and $\mathbf{v}_2$. The question is: what are their velocities after the collision? It is assumed that the collision takes place so quickly that the only thing that needs to be considered is the collision event itself and that other influences from external force fields can be regarded as constant. Calling the velocities of the two bodies after the collision $\mathbf{u}_1$ and $\mathbf{u}_2$ and assuming no other external forces, the sum of forces on the bodies is zero and therefore conservation of momentum is applied

$$m_1\mathbf{v}_1 + m_2\mathbf{v}_2 = m_1\mathbf{u}_1 + m_2\mathbf{u}_2 \qquad (1.40)$$

Here are three equations with six unknowns. Further information can be derived from the conservation of angular momentum. To investigate these, use a special coordinate frame. The incoming velocities define a plane; take this to be the (x, y)-plane, so that the z-components of the incoming velocities are zero. The x-component of the angular momentum is now zero, because the z-coordinates are zero. It follows that the outgoing angular momentum also has a zero x-component. It is then seen that the whole process lies in a plane, which is the x, y-plane. Trivially, the same holds for the y-component, so this is not a new equation. The only angular momentum component that is relevant is now the z-component.

The count of unknowns and equations comes to: four unknowns, one angular momentum component and two momentum components. So there is one equation missing. That one will have to be the conservation or loss of energy. There are two possibilities: energy is conserved in the collision; this case is called an *elastic collision* and is relevant to very hard or very elastic bodies, such as steel spheres, rubber balls and certain atomic-scale collisions. If energy is not conserved, then a specification has to be made of how energy is lost; this is an *inelastic collision*. The phrase 'the energy is lost' is somewhat misleading. The energy has not evaporated into thin air. It would be more accurate to say that mechanical energy has been converted into heat, for example, radiation, or the internal energy of the participating bodies has gone up, that is, they have become warmer.

As before, with the bouncing ball example, a coefficient of restitution needs to be specified, which is a material parameter, characteristic of the materials of which the (surfaces) of the materials are made. It is essentially a phenomenological theory and there are various versions. For point masses the coefficient can be conveniently defined as the ratio of the root of the sums of kinetic energies of outgoing to incoming objects

$$e = \sqrt{\frac{\frac{1}{2}m_1u_1^2 + \frac{1}{2}m_2u_2^2}{\frac{1}{2}m_1v_1^2 + \frac{1}{2}m_2v_2^2}} \qquad (1.41)$$

Another definition of the coefficient of restitution is via the specification of the ratio of the normal outgoing and incoming velocities. For collisions that are head-on and take place in one dimension, the two definitions coincide.

When $e = 1$ the collision is elastic and the mechanical energy is conserved. During inelastic collisions mechanical energy is lost (and transferred to some other form) and $e < 1$. In a totally inelastic collision $e = 0$ and the two bodies stick together after the encounter.

For finite-sized bodies the description of the collision rheology depends on a number of phenomenological properties of the surfaces, such as Coulomb friction. Also, the rotation of the bodies needs to be taken into account. As yet this is a field that is still in development and the subject of muscular discussions in the engineering literature.

1.4.5.2   Disintegrating Bodies

When bodies disintegrate, in essence the opposite of an inelastic collision takes place. Here it will be assumed that the disintegration occurs progressively, which is a scenario in which small amounts of matter are expelled from a principal body more or less continuously. Examples of such processes are an asteroid disintegrating as it approaches a large planet and fluid or solid grains leaking from a moving object. The process may be quite deliberate in, for instance, rocket propulsion.

A body with time-dependent mass $m(t)$ has velocity v. The mass loss will be described as a series of small distinct ejections. So, in a time Δt an amount Δm is expelled at a velocity $\mathbf{v}_e$. $\mathbf{v}_e$ has a component in the direction of motion of the main body, which has a magnitude u_e (Fig. 1.8). The conservation of momentum, evaluated before and after the ejection, is written down under the assumption that the

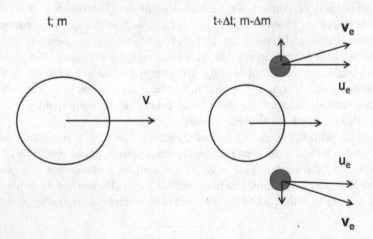

Fig. 1.8 Increment of mass loss

ejection takes place in such a way that the effective velocity of the ejected matter has no mean component in the direction of motion of the main body, so that $< \mathbf{v}_e >= u_e$ (this does not mean that $< (\mathbf{v}_e)^2 >= u_e^2$).

$$m(t)v = (m(t) - \Delta m)v(t + \Delta t) + \Delta m u_e \qquad (1.42)$$

Up the first order in the variations and dividing by Δm

$$- v(t) + m(t)\frac{\Delta v}{\Delta m} + u_e = 0 \qquad (1.43)$$

It follows that

$$\frac{\Delta v}{\Delta m} = \frac{v(t) - u_e}{m} \qquad (1.44)$$

Energy conservation cannot be satisfied without adding an external energy term Q, which may be specified per unit ejected mass. Another element that enters in is a representation of the mass being ejected sideways, which carries an amount of kinetic energy $\frac{1}{2}\Delta m D$, where D is the mean square component of the velocity normal to the mean motion. The conservation of energy then takes the form

$$\frac{1}{2}mv(t)^2 + Q\Delta m = \frac{1}{2}(m - \Delta m)(v(t) + \Delta v)^2 + \frac{1}{2}\Delta m(u_e^2 + D) \qquad (1.45)$$

Again, taking terms up to first order in the variations gives the result

$$Q = \frac{1}{2}v^2 - vu_e + \frac{1}{2}(u_e^2 + D) \qquad (1.46)$$

If no matter is ejected $u_e = v$ and $D = 0$; in that case $Q = 0$. Otherwise,

$$Q = \frac{1}{2}(v_e - u)^2 + \frac{1}{2}D \qquad (1.47)$$

It follows that in order to expel mass energy needs to be supplied. The more energy is supplied, the greater the difference in velocity between the ejected matter and the moving object.

An example is the case of rocket propulsion, where the energy is supplied from the chemical energy stored in the fuel, which is released on burning. Another well-known example is a balloon, which is blown up and then let go; the elastic energy stored in the wall of the balloon is released and the air can be expelled.

1.4.5.3 The Harmonic Oscillator with Mass Loss

The energy stored in a harmonic oscillator can also be used to eject matter. A special case would be the one in which the matter is expelled at right angles to the motion, so that the only term that matters is D. (Again, it is assumed that the expulsion is symmetric.) A practical realisation of such a system is a plastic water bottle, which has a small hole in it, so that water can seep out. The bottle is suspended from a string and set in pendular motion. The expulsion in this case is not symmetric, but the momentum transferred in the direction of the string is easily absorbed by the tension in the string.

The energy of the oscillator is

$$E = \frac{1}{2}m(t)\dot{x}^2 + \frac{1}{2}kx^2 \qquad (1.48)$$

As energy is lost a damping term $\gamma\dot{x}$ has to be inserted into the equation of motion. In Eq. (1.47) Q represents the energy requirement per unit expelled mass. So, multiplying by Δm and dividing by Δt give the rate of total energy required. Thus, $\dot{E}$ equals $\frac{1}{2}\dot{m}D$. The rate of change of the energy is

$$\dot{E} = \frac{1}{2}\dot{m}\dot{x}^2 + \frac{1}{2}m(t)\frac{d}{dt}(\dot{x})^2 + \frac{1}{2}\frac{d}{dt}(kx^2) \qquad (1.49)$$

The equation of motion is multiplied with $\dot{x}$

$$\gamma(\dot{x})^2 + \dot{m}(\dot{x})^2 + m(t)\dot{x}\ddot{x} + k\dot{x}x = 0, \qquad (1.50)$$

which implies

$$\gamma(\dot{x})^2 + \dot{m}(\dot{x})^2 + m(t)\frac{1}{2}\frac{d}{dt}(\dot{x})^2 + \frac{1}{2}\frac{d}{dt}(kx^2) = 0 \qquad (1.51)$$

Substituting this in Eq. (1.49) leads to

$$\dot{E} = -\frac{1}{2}\dot{m}(\dot{x})^2 - \gamma(\dot{x})^2 = \frac{1}{2}\dot{m}D \qquad (1.52)$$

A form of the damping factor γ then follows

$$\gamma = -\dot{m}\left(\frac{1}{2} + \frac{1}{2}\frac{D}{(\dot{x})^2}\right) \qquad (1.53)$$

Now, in order to understand the result better make a modelling assumption: set the ratio $D/(\dot{x})^2$ equal to a constant value, which will be called ζ. This parameter gives a measure for how much energy is carried away by the ejected mass. The equation of motion takes the form

$$\frac{1}{2}\dot{m}\dot{x} - \frac{1}{2}\dot{m}\zeta\dot{x} + m\ddot{x} + kx = 0 \qquad (1.54)$$

It is not unreasonable to assume that the ejected mass is proportional to the available mass with proportionality constant β, so that $\dot{m} = -\beta m$ and therefore $m(t) = m_0 e^{-\beta t}$ (where m_0 is the mass at time $t = 0$). The equation of motion that results takes the form

$$-\frac{\beta}{2}(1 - \zeta)\dot{x} + \ddot{x} + \frac{k}{m_0}e^{\beta t}x = 0 \qquad (1.55)$$

The first derivative can be eliminated by setting $x(t) = u(t)e^{\beta(1-\zeta)t/4}$; the function $u(t)$ satisfies

$$\ddot{u} - \left(\frac{\beta^2}{16}(1 - \zeta)^2 - \frac{k}{m_0}e^{\beta t}\right)u = 0 \qquad (1.56)$$

A differential equation of this form has the general solution

$$u(t) = A_1 J_{\frac{|1-\zeta|}{2}}\left(\frac{2}{\beta}\sqrt{\frac{k}{m_0}}e^{\beta t/2}\right) + A_2 Y_{\frac{|1-\zeta|}{2}}\left(\frac{2}{\beta}\sqrt{\frac{k}{m_0}}e^{\beta t/2}\right), \qquad (1.57)$$

where $J_\nu(z)$ and $Y_\nu(z)$ are Bessel functions of the first and second kind of order ν and A_1 and A_2 are constants which are chosen to fit initial conditions. (The properties of Bessel functions are elucidated in Sect. A.8.3 and in Abramowitz and Stegun (1972), Boas (1983).) The solution for $x(t)$ is now

$$x(t) = e^{\beta(1-\zeta)t/4}\left[A_1 J_{\frac{|1-\zeta|}{2}}\left(\frac{2}{\beta}\sqrt{\frac{k}{m_0}}e^{\beta t/2}\right) + A_2 Y_{\frac{|1-\zeta|}{2}}\left(\frac{2}{\beta}\sqrt{\frac{k}{m_0}}e^{\beta t/2}\right)\right] \qquad (1.58)$$

The character of this solution is easily investigated by inspecting the asymptotic behaviour for $t \to \infty$. The Bessel functions behave as (see again Sect. A.8.3)

$$J_\nu(z) \to \sqrt{\frac{2}{\pi z}}\left[\cos(z - \frac{\nu\pi}{2} - \frac{\pi}{4}) + O(|z|^{-1})\right]; \qquad (1.59)$$

$$Y_\nu(z) \to \sqrt{\frac{2}{\pi z}}\left[\sin(z - \frac{\nu\pi}{2} - \frac{\pi}{4}) + O(|z|^{-1})\right] \qquad (1.60)$$

In terms of the time-dependence of the displacement in this limit the solution behaves as

$$x(t) \to e^{-\zeta t/4}\left[\tilde{A}_1 \cos\left(\frac{2}{\beta}\sqrt{\frac{k}{m_0}}e^{\beta t/2}\right) + \tilde{A}_2 \sin\left(\frac{2}{\beta}\sqrt{\frac{k}{m_0}}e^{\beta t/2}\right)\right], \qquad (1.61)$$

where $\tilde{A}_1$ and $\tilde{A}_2$ are two constants to fit suitably defined conditions.

The character of this solution is an oscillation, as expected, with ever-increasing frequency and an exponential amplitude decay, which goes as $e^{-\zeta t/4}$. Thus when ζ is zero there is no exponential decay, but in that case no mass is ejected (β must then also be zero and the ordinary, undamped harmonic oscillator solution results). For non-zero ζ there is exponential decay, so *mass loss leads to damping*.

Note that this whole analysis only considers a purely mechanical balance; there may well be extra damping due to heat losses. To accommodate those a more extensive analysis is required, but also more assumptions have to be made concerning the details of the process of heat loss.

References

Abramowitz M, Stegun A (1972) Handbook of mathematical functions. Dover, New York

Boas ML (1983) Mathematical methods in the physical sciences. Wiley, New York

Kibble TWB (1985) Classical mechanics, 3rd edn. Longman Scientific and Technical, Harlow

Koenders MAC (2020) The physics of the deformation of densely packed granular materials. World Scientific, Singapore

Newton I (1687) Philosophi naturalis principia mathematica. The Royal Society (printer Joseph Streater), London

Newton I (1995) The principia (tr Motte, A.). Prometheus Books, New York

Chapter 2
Newtonian Mechanics Reformulated

Abstract Newtonian mechanics needs to be reformulated to make it both more applicable in more cases and also to make it amenable to a form in which the laws of mechanics can be obtained from a principle of least action. The task of rewriting Newton's laws of motion in a rotating frame leads to the introduction of fictitious forces, notable the Coriolis force and the centrifugal force. A new mechanical principle is then introduced—the principle of d'Alembert—to deal with constrained systems and reaction forces. The application of this principle enables the mechanical laws to be reformulated and leads to Lagrange's equations of the first kind. Then, using generalised coordinates and generalised momenta, a more efficient formulation of the equations of motion is obtained via Lagrange's equations of the second kind. Conservation laws (constants of motion) are derived. Examples of the use of the Lagrange equations are shown, including the Bateman method of describing damping.

2.1 Why Mechanics Needs to Be Reformulated

There is no doubt that Newton's equations work very well in one-dimensional problems. Newton's second law is essentially phrased in vector terms, based as it is on the inertial principle, which prescribes straight lines, attached to the fixed, most distant, stars. If, however, it is *natural* (or convenient/elegant) for a system to be formulated in a coordinate system that rotates with respect to the fixed stars then special measures have to be taken. This is to compensate for the fact that the inertial principle will want to force an object in a straight line, while a body that is attached to a rotating system does not follow that path. To represent the tendency for a body to follow a straight line with respect to the fixed stars, so-called *fictitious forces* have to be introduced. For example, in a car that goes fast round the corner, passengers will feel a sideways force. The passengers would like to follow the fixed stars and go on in a straight line, but the car (their coordinate frame) forces them on a bended curve.

Another reformulation that is required is concerned with rigid elements. It could be argued that there are no completely rigid bodies in physics, nevertheless for practical purposes it is often convenient—and a good approximation—to regard them as such. For example, for a pendulum with a mass point at the tip of a rigid arm, the motion of the mass point is constrained by the arm and it clearly cannot follow a straight line. The arm exerts a force on the mass point; such forces are called *reaction forces*. They occur generally in constrained motion.

Furthermore, it may be asked whether a formulation of mechanics cannot be achieved by taking a measure of the energy as a starting point. In the first instance this would be relevant in cases in which energy is conserved, such as for most astronomical calculations and for motions that take place on the (sub-)molecular scale. Damping is the result of collisions with molecules of a surrounding gas or fluid and is generally velocity dependent (not necessarily linearly). Reformulating mechanics on the basis of energy functions will produce a very elegant way of thinking about physical processes and is applicable far beyond classical mechanics. There is then ample scope to investigate a reformulation of Newtonian mechanics, not to invalidate it, but to make it more applicable to specific cases and to extend the understanding of physics more generally.

2.2   Newton's Second Law in a Rotating Coordinate Frame

A rotating frame is described by defining an axis around which the system spins. The axis is set with respect to the fixed stars. In the fixed-star system, position vectors are called $\mathbf{r}$; in the rotating system, they are called $\mathbf{r}'$. The transformation from the fixed system to the rotating system is accomplished by defining an angular velocity $|\omega|$. A vector is defined that points in the direction of the axis and has magnitude $|\omega|$; the vector is denoted as ω, defining both the direction of rotation and its magnitude. The direction of the vector fits the rotation as a right-handed screw. If now the rate of change in the rotating frame is to be evaluated, the time-dependence of the frame must be kept in mind. For illustration a sketch is presented in Fig. 2.1.

Fig. 2.1 Rate of change of a unit vector in a rotating coordinate frame

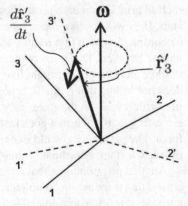

First the rates of change of the unit vectors of the rotating frame are evaluated. For each unit vector $\hat{\mathbf{r}}'_i$, where i is 1, 2 or 3, it can be read from Fig. 2.1 (illustrated for $i = 3$) that

$$\frac{d\hat{\mathbf{r}}'_i}{dt} = \omega \times \hat{\mathbf{r}}'_i \tag{2.1}$$

An arbitrary vector $\mathbf{A}$ has two representations: one in the frame fixed to the stars, and the other in the rotating frame. The former is $\mathbf{A} = A_1\hat{\mathbf{r}}_1 + A_2\hat{\mathbf{r}}_2 + A_3\hat{\mathbf{r}}_3$, the latter $\mathbf{A} = A'_1\hat{\mathbf{r}}'_1 + A'_2\hat{\mathbf{r}}'_2 + A'_3\hat{\mathbf{r}}'_3$. The rate of change is d/dt in the fixed system and is d'/dt as measured in the rotating system. The connection between the two must take account of the fact that the rotating coordinate system itself also has a rate of change. So, differentiating

$$\frac{d\mathbf{A}}{dt} = \frac{d'\mathbf{A}}{dt} + \sum_{i=1}^{3} A'_i \frac{d'\hat{\mathbf{r}}'_i}{dt} \tag{2.2}$$

The time derivative of the unit vectors is known, therefore

$$\frac{d\mathbf{A}}{dt} = \frac{d'\mathbf{A}}{dt} + \sum_{i=1}^{3} A'_i \omega \times \hat{\mathbf{r}}'_i = \frac{d'\mathbf{A}}{dt} + \omega \times \mathbf{A} \tag{2.3}$$

In this way the time dependency of the rotating frame, as viewed from the fixed frame, is accounted for; d'/dt is called the *co-moving derivative*.

The first vector to apply this to is the position vector $\mathbf{r}$. The rate of change is

$$\frac{d\mathbf{r}}{dt} = \frac{d'\mathbf{r}}{dt} + \omega \times \mathbf{r} \tag{2.4}$$

or

$$\mathbf{v} = \mathbf{v}' + \omega \times \mathbf{r} \tag{2.5}$$

The acceleration is found by applying the same rule again, so, assuming that ω is constant,

$$\frac{d\mathbf{v}}{dt} = \frac{d'}{dt}\frac{d'\mathbf{r}}{dt} + \omega \times \frac{d'\mathbf{r}}{dt} + \frac{d'}{dt}(\omega \times \mathbf{r}) + \omega \times (\omega \times \mathbf{r}) = \mathbf{a}' + 2\omega \times \mathbf{v}' + \omega \times (\omega \times \mathbf{r}) \tag{2.6}$$

Newton's second law can then be used in a rotating frame, but the two extra terms have to be inserted into the acceleration to make it appropriate to the inertial frame (in which Newton's first law holds). The term $\omega \times (\omega \times \mathbf{r})$ is called the *centripetal acceleration*. The same term with a minus sign is the *centrifugal acceleration*. This is the acceleration that is felt when a body is forced on to a curved path. The term $-2\omega \times \mathbf{v}'$ is called the *Coriolis acceleration* (Gaspard-Gustave de Coriolis 1792–1843). The effects of this term in the acceleration and the fictitious forces that arise from them are generally less well known.

The order of magnitude of these fictitious (sometimes called 'apparent') forces is easily ascertained. An obvious rotating system is the Earth, its radius is 6371 km and the value of $|\omega|$ is $2\pi/(24 \times 3600)$. The centrifugal acceleration at the equator is therefore $0.0337\,\mathrm{ms}^{-2}$, which is negligible compared to the value of g of approximately $9.8\,\mathrm{ms}^{-2}$. However, for a spinning disc of diameter 0.5 m, that makes 1 revolution per second, the centrifugal acceleration is about equal to the gravitational acceleration. So, for example, for a spinning car wheel the centrifugal force is important. Similarly, a body on the rim of a fairground ride with a diameter of 20 m, doing 1 revolution every 10 seconds, experiences an acceleration of approximately $0.4g$, which is very noticeable (and the whole point of the ride).

To estimate the Coriolis force a velocity has to be specified. Again, taking the Earth as an example, let a body move North on the equator at a velocity of $v' = 100\,\mathrm{ms}^{-1}$. The acceleration a_C it experiences can be estimated and is about $0.014\,\mathrm{ms}^{-2}$. Thus, the deflection of the body after having travelled some 100 km (S) is roughly 7 km ($\frac{1}{2}a_C S^2/v'^2$). This is small and it shows that the effects of the Coriolis force on Earth are only felt over very large distances, typically, in meteorological and oceanographic applications. Cyclonic and anti-cyclonic weather patterns at temperate latitudes around low and high pressure areas can be explained by the Coriolis force. Note that because of the presence of the outer product in the Coriolis force, its effects work in an opposite direction in the southern hemisphere to the ones in the northern hemisphere. The wind associated with an anti-cyclone in the northern hemisphere goes around clockwise (as viewed from the top), while in the southern hemisphere it follows an anti-clockwise direction. The smallness of the Coriolis force due to the rotation of the Earth makes a nonsense of the assertion that bath tubs empty in a clockwise vortex on the northern hemisphere and anti-clockwise on the southern hemisphere. If there is an effect it must be due to the fact that baths are shaped differently in either hemisphere. The connection with the Coriolis force is purely mythological. On the smaller scale, walking at $0.5\,\mathrm{ms}^{-1}$ across a fairground attraction that spins at 1 revolution in every 10 seconds one experiences a Coriolis acceleration of some $0.6\,\mathrm{ms}^{-2}$, much less than the centrifugal force or the gravitation force, but maybe just noticeable.

2.3 Systems with Constraints, d'Alembert's Principle

This section is concerned with systems of mass points that are somehow constrained in their movement. Examples are a pendulum with a rigid arm, a train on rails, a pair of dumbbells with a rigid connection between them, etc. The constraint will in general exert a force; these are called *reaction forces*. In principle—merely applying Newton's laws—there is no way of knowing what they are. Take the (somewhat idealised) example of the pendulum with a rigid arm of negligible mass. The mass point at the tip of the arm feels a force from the arm, but the only known force in the system is the gravity force on the mass point. Also, in this case, the motion of the pendulum is perfectly described by the one coordinate (for example, the angle of the deviation), but for two-dimensional motion Newton's equation has two components.

This is generally the case in constrained systems: the number of degrees of freedom is smaller than the number of coordinates of the problem.

When the constraints are functions of the coordinates and the time only (that is, *not* the velocities), the system is said to be *holonomic*. When they are independent of the time they are said to be *scleronomic* (*rheonomic* when they depend explicitly on the time). Therefore, the constraints may be written as functions. There are n coordinates and p constraints; it is an obvious necessity that $p < n$. The constraints take the functional form

$$G_j(q_1, q_2, \ldots q_n, t) = 0; \quad j = 1 \ldots p \qquad (2.7)$$

For example, a mass point constrained to move on a sphere with radius R has coordinates $q_1 = r_1, q_2 = r_2, q_3 = r_3$ and the scleronomic constraint is $G_1 = r_1^2 + r_2^2 + r_3^2 - R^2 = 0$. A mass point on an inflating sphere is described by the constraint $G_1 = r_1^2 + r_2^2 + r_3^2 - R(t)^2 = 0$; this is a rheonomic constraint. Parenthetically it is noted that for solid bodies the angular orientation must also be specified, so the number of coordinates is six: three position coordinates and three angles. Constraints are then imposed in terms of these six coordinates. A system of two solid bodies has twelve coordinates: six positions and six angles and so on. For more complex systems, the bookkeeping becomes quite daunting. For the moment the discussion will be limited to mass points. A detailed account of the motion of solid bodies is found in Kibble (1985).

In terms of the physics the constraint is realised via reaction forces. Newton's second law needs to be supplemented with reaction forces $\mathbf{R}$. For n coordinates there are n equations with n extra reaction forces

$$m_\nu \ddot{\mathbf{r}}_\nu = \mathbf{F}_\nu + \mathbf{R}_\nu, \quad \nu = 1 \ldots n \qquad (2.8)$$

Here $\mathbf{F}_\nu$ represent the known (external) forces. There are now $n + p$ equations $2n$ unknowns ($\mathbf{r}_\nu$ and $\mathbf{R}_\nu$), so information is lacking. A new principle is required to supply it. The principle that achieves this is *d'Alembert's principle* and in words it is phrased as follows. *For a small (infinitesimal) virtual displacement the reaction forces do not do any work.* There is no proof for the correctness of the principle. It is a new, independent postulate in mechanics, though intuitively it must be correct.

To see how the principle works consider the case of a rigid pendulum with a massless arm of length L and a mass point of mass m at the tip. Let the pendulum move in two dimensions, then there are two coordinates, which are called x and y; the deflection angle is called ϕ. The acceleration due to gravity is g (Fig. 2.2).

The second law is

$$m\ddot{x} = R_x(t); \quad m\ddot{y} = R_y(t) + mg \qquad (2.9)$$

The constraint is

$$x^2 + y^2 - L^2 = 0 \qquad (2.10)$$

Fig. 2.2 Sketch of the
configuration of a pendulum

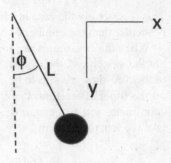

There are four unknowns. Use the constraint to express $x(t)$ and $y(t)$ in terms of ϕ; in this way one unknown is eliminated

$$\dot{x} = L\cos(\phi)\dot{\phi}; \quad \ddot{x} = -L\sin(\phi)\dot{\phi}^2 + L\cos(\phi)\ddot{\phi} \tag{2.11}$$

$$\dot{y} = -L\sin(\phi)\dot{\phi}; \quad \ddot{y} = -L\cos(\phi)\dot{\phi}^2 - L\sin(\phi)\ddot{\phi} \tag{2.12}$$

to give

$$-mL\sin(\phi)\dot{\phi}^2 + mL\cos(\phi)\ddot{\phi} = R_x; \quad -mL\cos(\phi)\dot{\phi}^2 - mL\sin(\phi)\ddot{\phi} = R_y + mg \tag{2.13}$$

Now enforce d'Alembert's principle. The reaction force does not do any work and therefore the direction of $\mathbf{R}$ is perpendicular to the motion, in other words $R_x = R(t)\sin(\phi)$ and $R_y = R(t)\cos(\phi)$. $R(t)$ can then be eliminated from the set of equations, to yield

$$\ddot{\phi} + \frac{g}{L}\sin(\phi) = 0 \tag{2.14}$$

Two observations are made. Firstly, for small values of ϕ, $\sin(\phi)$ may be approximated to ϕ and then the equation for the harmonic oscillator is obtained. This is a linear equation. It has an oscillating solution with circular frequency $\omega = \sqrt{g/L}$. Note that this is independent of the mass; the period of oscillation depends on the length of the arm. Secondly, multiplying with $\dot{\phi}$ gives an integral of the form $E = \frac{1}{2}\dot{\phi}^2 - \frac{g}{L}\cos(\phi)$.

Depending on E the portrait in phase space, in which $\dot{\phi}$ is plotted as a function of ϕ, has characteristic features. When the energy is small, an elliptical orbit in phase space is followed. For larger energies the amplitude also becomes larger. When the energy is large enough the pendulum can carry out an orbit that goes round and round and never reverses. Examples of these phase portraits are illustrated in Fig. 2.3.

Using the integral E the reaction force can be expressed in the angle ϕ

$$R(t) = -2mLE - 3mg\cos(\phi) \tag{2.15}$$

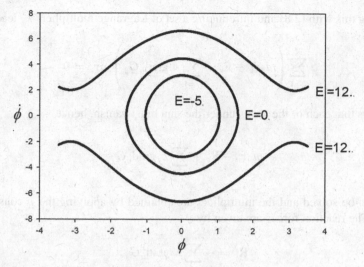

Fig. 2.3 Phase portraits of a pendulum for different values of E

So, this example shows that d'Alembert's principle can indeed be used to calculate the motion of a system and find the reaction forces. However, as noted above, d'Alembert's principle cannot be proven; it is an assumed (though highly plausible) element in the theory of mechanics.

2.4 Lagrange's Equations of the First Kind

The Lagrange equations of the first kind also employ d'Alembert's principle, but permit a solution of the second law by using Lagrange multipliers (see Sect. A.4). For a system of N particles there are $n = 3N$ Cartesian coordinates and hence n components of the second law. For a constrained system there are obviously no solutions, but the constraints can be applied via d'Alembert's principle. The constraints take the form

$$G_j(r_1, r_2, ...r_n, t) = 0, \quad j = 1...p \tag{2.16}$$

A variation of the constraints under a virtual displacement set of vectors is

$$\sum_\nu \mathrm{grad}_\nu G_j.\delta\mathbf{r}_\nu = 0 \tag{2.17}$$

The reaction forces do not do any work so

$$\sum_\nu \mathbf{R}_\nu.\delta\mathbf{r}_\nu = 0 \tag{2.18}$$

Applying this with (2.8) and introducing a set of Lagrange multipliers λ_j leads to

$$\sum_{\nu=1}^{N} \left(m_\nu \ddot{\mathbf{r}}_\nu - \mathbf{F}_\nu + \sum_{j=1}^{p} \lambda_j \operatorname{grad}_\nu G_j \right) . \delta \mathbf{r}_\nu = 0 \qquad (2.19)$$

It follows that each of the terms under the sum must vanish, hence

$$m_\nu \ddot{\mathbf{r}}_\nu - \mathbf{F}_\nu + \sum_{j=1}^{p} \lambda_j \operatorname{grad}_\nu G_j \qquad (2.20)$$

They can be solved and the multipliers are obtained by applying the p constraints (2.16). The reaction forces are given by

$$\mathbf{R}_\nu = -\sum_j \lambda_j \operatorname{grad}_\nu G_j \qquad (2.21)$$

Note that the multipliers may be functions of time.

These equations will serve as a basis for the development of a more general formulation of mechanics.

2.5 Lagrange Equations of the Second Kind

The example of the pendulum has demonstrated that while there are two coordinates (x and y), the system can be described with only one variable ϕ. The question then is: if one is not interested in the reaction forces, is it not possible to phrase the second law directly in this one variable? When, more generally, there are n coordinates and p relations between them that fix the constraints, what is wanted is a form of mechanics that works with $n - p$ so-called *generalised coordinates*. These will be denoted by q_k, where $k = 1, 2, ...n - p$. There are therefore as many generalised coordinates as there are *degrees of freedom*. The time derivative of q_k is the *generalised velocity* $\dot{q}_k$. In order to get from the Cartesian coordinate second law to a second law framed in terms of generalised coordinates, the constraints have to somehow be eliminated.

The Cartesian coordinates can, of course, be expressed in the generalised coordinates (and the time t): $r_1 = r_1(q_1, q_2...q_{n-p}, t)$; $r_2 = r_2(q_1, q_2...q_{n-p}, t).....r_n = r_n(q_1, q_2...q_{n-p}, t)$. The constraints $G_j = 0$ in terms of the generalised coordinates are identically zero

$$G_j \left(r_1(q_k), r_2(q_k), ..., r_n(q_k), t \right) = 0 \qquad (2.22)$$

It follows that

$$\frac{\partial G_j}{\partial q_m} = \sum_{\nu=1}^{N} \left(\text{grad}_\nu G_j \cdot \frac{\partial \mathbf{r}_\nu}{\partial q_m} \right) \tag{2.23}$$

Now take the inner product of the Lagrange equations of the first kind with $\partial \mathbf{r}/\partial q_m$ and sum over all the N mass points ν. Then using (2.23) it is seen that all the terms with the Lagrange multipliers λ_j disappear.

$$\sum_{\nu=1}^{N} \left(m_\nu \ddot{\mathbf{r}}_\nu \cdot \frac{\partial \mathbf{r}_\nu}{\partial q_m} - \mathbf{F}_\nu \cdot \frac{\partial \mathbf{r}_\nu}{\partial q_m} \right) \tag{2.24}$$

for $m = 1, 2....n - p$. Furthermore, the generalised velocities are

$$\dot{\mathbf{r}}_\nu = \sum_{m=1}^{n-p} \frac{\partial \mathbf{r}_\nu}{\partial q_m} \dot{q}_m + \frac{\mathbf{r}_\nu}{\partial t} \tag{2.25}$$

The Cartesian velocities are therefore linear functions of the generalised velocities; it follows that

$$\left(\frac{\partial \dot{\mathbf{r}}_\nu}{\partial \dot{q}_m} \right)_{q_k} = \frac{\partial \mathbf{r}_\nu}{\partial q_m} \tag{2.26}$$

From which follows the identity

$$\ddot{\mathbf{r}}_\nu \cdot \frac{\partial \dot{\mathbf{r}}_\nu}{\partial \dot{q}_m} = \frac{d}{dt} \left(\dot{\mathbf{r}}_\nu \cdot \frac{\partial \dot{\mathbf{r}}_\nu}{\partial \dot{q}_m} \right) - \dot{\mathbf{r}}_\nu \frac{d}{dt} \left(\frac{\partial \mathbf{r}_\nu}{\partial q_m} \right) \tag{2.27}$$

The last term in brackets is further developed

$$\frac{d}{dt} \left(\frac{\partial \mathbf{r}_\nu}{\partial q_m} \right) = \frac{\partial^2 \mathbf{r}_\nu}{\partial q_m \partial t} + \sum_{k=1}^{n-p} \frac{\partial^2 \mathbf{r}_\nu}{\partial q_m \partial q_k} \dot{q}_k = \frac{\partial}{\partial q_m} \left(\frac{\mathbf{r}_\nu}{\partial t} + \sum_{k=1}^{n-p} \frac{\partial \mathbf{r}_\nu}{\partial q_k} \dot{q}_k \right) = \frac{\partial \dot{\mathbf{r}}_\nu}{\partial q_m}$$
$$\tag{2.28}$$

The result is substituted in (2.24) to give

$$\sum_{\nu=1}^{N} m_\nu \frac{d}{dt} \left(\dot{\mathbf{r}}_\nu \cdot \frac{\partial \dot{\mathbf{r}}_\nu}{\partial \dot{q}_m} \right) - \sum_{\nu=1}^{N} m_\nu \dot{\mathbf{r}}_\nu \cdot \frac{\partial \dot{\mathbf{r}}_\nu}{\partial \dot{q}_m} = \sum_{\nu=1}^{N} \mathbf{F}_\nu \cdot \frac{\partial \dot{\mathbf{r}}_\nu}{\partial \dot{q}_m} \tag{2.29}$$

The kinetic energy $T = \sum_{\nu=1}^{N} \frac{1}{2} m_\nu \dot{\mathbf{r}}_\nu^2$ is easily recognised and therefore the equation may be recast in the form

$$\frac{d}{dt} \left(\frac{\partial T}{\partial \dot{q}_m} \right) - \frac{\partial T}{\partial q_m} = \sum_{\nu=1}^{N} \mathbf{F}_\nu \cdot \frac{\partial \mathbf{r}_\nu}{\partial q_m} \tag{2.30}$$

The right-hand side of this equation is called the *generalised force Q_m*. The resulting equation can be summarised as

$$\frac{d}{dt}\left(\frac{\partial T}{\partial \dot{q}_m}\right) - \frac{\partial T}{\partial q_m} = Q_m \tag{2.31}$$

Equation (2.31) is the *Lagrange equation of the second kind*. It formulates the mechanics directly in the generalised coordinates. No constraint forms are necessary and there is great freedom in how to choose these coordinates, which control all the degrees of freedom of the system. What is not possible though is to obtain the reaction forces; for these the Lagrange equations of the first kind are required.

A further development may be added if the forces are associated with a potential. As observed, the Cartesian coordinates can be expressed in the generalised coordinates. Therefore, the potential energy V may be expressed in the generalised coordinates

$$Q_m = -\frac{\partial V}{\partial q_m} \tag{2.32}$$

Now, the Lagrange function is introduced, also known as the *Lagrangian*, which is a function of the generalised coordinates and generalised velocities and denoted by $\pounds(q_m, \dot{q}_m)$

$$\pounds = T - V \tag{2.33}$$

Then the Lagrange equations (of the second kind) are

$$\frac{d}{dt}\left(\frac{\partial \pounds}{\partial \dot{q}_m}\right) - \frac{\partial \pounds}{\partial q_m} = 0 \tag{2.34}$$

The solutions of these equations are obtained under the specification of initial conditions in terms of $q_m(0)$ and $\dot{q}_m(0)$.

The Lagrange equations have been derived using the following assumptions:
1. The system is holonomic.
2. d'Alembert's principle is valid.
3. There exists a potential that depends on the generalised coordinates only, but *not* on the generalised velocities.

In addition to generalised coordinates, generalised velocities and generalised forces, the *generalised momentum* is defined as

$$p_m = \frac{\partial \pounds}{\partial \dot{q}_m} \tag{2.35}$$

A number of examples of the use of the Lagrange equations will be given below. Their applications reach well beyond traditional classical mechanics; in some sense it could be said that modern physics starts here. It is difficult to overstate the importance of Lagrange's approach to mechanics. His own work (Lagrange (1789)), published in two volumes a hundred years after the *Principia*, became the standard work on mechanics for next century.

2.6 Energy Integral and Cyclic Coordinates

In this section the Lagrange equations are deployed to obtain constants of motion. Some of these are already known from Chapter 1, indicating a classical limit. The Lagrangian formalism, however, enables a direct route to a class of conservation laws in a very transparent manner.

2.6.1 Jacobi's Integral and the Hamiltonian

The total time derivative of the Lagrangian is

$$\frac{d£}{dt} = \frac{\partial £}{\partial t} + \sum_k \frac{\partial £}{\partial q_k}\dot{q}_k + \sum_k \frac{\partial £}{\partial \dot{q}_k}\ddot{q}_k \tag{2.36}$$

Substituting the Lagrange equations then leads to

$$\frac{d£}{dt} = \frac{\partial £}{\partial t} + \frac{d}{dt}\sum_k \left(\frac{\partial £}{\partial \dot{q}_k}\dot{q}_k\right) \tag{2.37}$$

Consider the special case that the Lagrangian does not depend explicitly on the time, $\partial £/\partial t = 0$, then it follows that

$$\sum_k \left(\frac{\partial £}{\partial \dot{q}_k}\dot{q}_k\right) - £ = \tilde{H}(q_k, \dot{q}_k) = \text{constant} \tag{2.38}$$

This is *Jacobi's integral of motion*. It is an example of a conserved quantity.

A slightly more restrictive case is the one in which the kinetic energy is a quadratic function of the generalised coordinates. In that case the product under the sum in the Jacobi integral is just equal to twice the kinetic energy of the system and the integral becomes $2T - £ = T + V$ and therefore $\tilde{H}$ is the total energy, which is then a constant of motion.

The generalised velocities can be eliminated and expressed in the generalised momenta; then $\tilde{H}(q_k, \dot{q}_k) = H(p_k, q_k)$. This function is called the *Hamiltonian*. In the sections that follow it will be shown that the Hamiltonian can be employed for the purpose of yet another reformulation of classical mechanics. All these reformulations will reappear in other branches of physics and extensions of classical mechanics.

2.6.2   Cyclic Coordinates

A further search for conserved quantities leads to a rather simple result. A *cyclic coordinate* is defined as a generalised coordinate that is *absent from the Lagrangian*. In that case the associated momentum is conserved, as is easily seen

$$\frac{d}{dt}\left(\frac{\partial \pounds}{\partial \dot{q}_k}\right) = \frac{dp_k}{dt} = 0 \tag{2.39}$$

Hence p_k is a conserved quantity.

Below a number of examples of this mechanism will be given. It is a very powerful principle in physics.

2.7   Examples of the Use of Lagrange's Equations

Some examples are fairly trivial, however in some other cases it would have been really difficult to get a result in any other way.

2.7.1   A Free Particle and a Particle in a Conservative Force Field

Take the 'ordinary' Cartesian coordinates x, y and z and a potential $V(x, y, z)$. The Lagrangian is

$$\pounds = \frac{1}{2}m(\dot{x}^2 + \dot{y}^2 + \dot{z}^2) - V(x, y, z) \tag{2.40}$$

When the particle is free $V = 0$ and the coordinates x, y and z are all cyclic. The associated momenta are $p_x = \partial\pounds/\partial\dot{x}$, $p_y = \partial\pounds/\partial\dot{y}$ and $p_z = \partial\pounds/\partial\dot{z}$ and these will be conserved quantities. They are, of course, the components of the ordinary, linear momentum vector and for this case it holds that

$$\mathbf{p} = m\dot{\mathbf{r}} = \text{constant} \tag{2.41}$$

When the particle is not free the 'ordinary' second law emerges

$$m\ddot{\mathbf{r}} + \text{grad } V = 0 \tag{2.42}$$

The energy $T + V$ is conserved, because the kinetic energy is homogeneously quadratic in the generalised velocities.

Equation (2.42) is entirely equivalent with Newton's second law when the force can be derived from a potential. Lagrangian mechanics thus converges to classical, Newtonian mechanics.

2.7.2 A Particle in a Plane in Two Dimensions

The coordinates that are chosen are the polar coordinates r, ϕ. The z direction is normal to the plane. The kinetic energy is easily obtained by differentiating the Cartesian coordinates with respect to time: $x = r \cos \phi$ and $y = r \sin \phi$, so

$$T = \frac{1}{2}m(\dot{r}^2 + r^2\dot{\phi}^2) \tag{2.43}$$

The Lagrangian is simply

$$\pounds = \frac{1}{2}m(\dot{r}^2 + r^2\dot{\phi}^2) - V(r, \phi) \tag{2.44}$$

It follows that the two Lagrange equations are

$$\frac{d}{dt}(m\dot{r}) - mr\dot{\phi}^2 + \frac{\partial V}{\partial r} = 0 \tag{2.45}$$

$$m\frac{d}{dt}(r^2\dot{\phi}) + \frac{\partial V}{\partial \phi} = 0 \tag{2.46}$$

A special case is when there is a central potential, that is, the potential only depends on the radial coordinate. Then ϕ is cyclic and the associated momentum $mr^2\dot{\phi}$ is conserved. This is just the z-component of the angular momentum.

The generalised forces are distinguished. $Q_r = -\partial V/\partial r$ is the radial force. $Q_\phi = -\partial V/\partial \phi$ is not a force as such, but the z component of the *force moment*.

2.7.3 The Equations of Motion in a Rotating Coordinate Frame

Cartesian coordinates are used. The transformation between the fixed-star coordinate frame x, y and the rotated frame x', y' is an *explicit* function of time

$$x = x' \cos \omega t - y' \sin \omega t; \quad y = x' \sin \omega t + y' \cos \omega t \tag{2.47}$$

The kinetic energy is

$$T = \frac{1}{2}m(\dot{x}^2 + \dot{y}^2) = \frac{1}{2}m(\dot{x}'^2 + \dot{y}'^2 + 2\omega x'\dot{y}' - 2\omega y'\dot{x}' + \omega^2 x'^2 + \omega^2 y'^2) \quad (2.48)$$

The two momentum components are

$$p_{x'} = \frac{\partial T}{\partial \dot{x}'} = m(\dot{x}' - \omega y') \quad p_{y'} = \frac{\partial T}{\partial \dot{y}'} = m(\dot{y}' + \omega x') \quad (2.49)$$

And the Lagrange equations are

$$m\ddot{x}' - 2m\omega\dot{y}' - m\omega^2 x' = -\frac{\partial V}{\partial x'} \quad (2.50)$$

$$m\ddot{y}' + 2m\omega\dot{x}' - m\omega^2 y' = -\frac{\partial V}{\partial y'} \quad (2.51)$$

Here the Coriolis force and the centrifugal force are easily identified, doing hardly any algebra!

2.7.4 The Double Pendulum, Small Oscillations

This is an example that would be very difficult to do without the framework of the Lagrange equations. In the limit of small oscillations the Lagrangian may be expanded in a Taylor series up to second order in the generalised coordinates and velocities, thereby ensuring that the equations of motion are *linear* differential equations.

For coordinates choose the angles with the vertical: ϕ_1 for mass m_1 and ϕ_2 for mass m_2. The two-dimensional Cartesian coordinates are $x_1 = l_1 \sin \phi_1$, $y_1 = l_1 \cos \phi_1$ and $x_2 = x_1 + l_2 \sin \phi_2$ $y_2 = y_1 + l_2 \cos \phi_2$—consult Fig. 2.4. The kinetic energy is

$$T = \frac{1}{2}m_1 l_1^2 \dot{\phi}_1^2 + \frac{1}{2}m_2 \left[l_1^2 \dot{\phi}_1^2 + l_2^2 \dot{\phi}_2^2 + 2l_1 l_2 \dot{\phi}_1 \dot{\phi}_2 \cos(\phi_1 - \phi_2) \right] \quad (2.52)$$

In the limit of small oscillations the term $\cos(\phi_1 - \phi_2)$ may be replaced by 1.

The potential energy is

$$V = (m_1 + m_2)gl_1(1 - \cos \phi_1) + m_2 g l_2(1 - \cos \phi_2) \quad (2.53)$$

and, again in the limit of small oscillations (up to second order in the ϕ_1 and ϕ_2), it is approximated as

$$V = \frac{1}{2}(m_1 + m_2)gl_1\phi_1^2 + \frac{1}{2}m_2 g l_2 \phi_2^2 \quad (2.54)$$

The Lagrange equations are

Fig. 2.4 Double pendulum

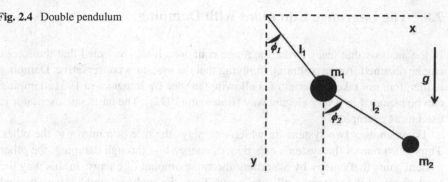

$$(m_1 + m_2)l_1\ddot{\phi}_1 + m_2l_2\ddot{\phi}_2 - (m_1 + m_2)g\phi_1 = 0 \qquad (2.55)$$

$$m_2l_1\ddot{\phi}_1 + m_2l_2\ddot{\phi}_2 - m_2g\phi_2 = 0 \qquad (2.56)$$

The eigen modes are found by setting

$$\phi_1(t) = \hat{\phi}_1 e^{i\omega t}; \quad \phi_2(t) = \hat{\phi}_2 e^{i\omega t} \qquad (2.57)$$

Leading to the two equations

$$\left(\omega^2(m_1 + m_2)l_1 + (m_1 + m_2)g\right)\hat{\phi}_1 + \omega^2 m_2 l_2 \hat{\phi}_2 = 0 \qquad (2.58)$$

$$\omega^2 m_2 l_1 \hat{\phi}_1 + \left(\omega^2 m_2 l_2 + m_2 g\right)\hat{\phi}_2 = 0 \qquad (2.59)$$

These two simultaneous equations have a solution for ω. By way of illustration take the special case that $l_1 = l_2 = l$, then the two solutions for ω^2 are

$$\omega^2 = \frac{g}{l}\left(\frac{m_1 + m_2}{m_1} \pm \sqrt{\frac{(m_1 + m_2)m_2}{m_1^2}}\right) \qquad (2.60)$$

The two solutions correspond to two possible modes of motion, given by their amplitude ratio.

$$\frac{\hat{\phi}_2}{\hat{\phi}_1} = \pm\sqrt{\frac{m_1 + m_2}{m_2}} \qquad (2.61)$$

These ratios depend on the mass ratio only. A positive value indicates that the motion of the two masses enhance each other, a negative one that they move against one another.

Other interesting examples of the use of the Lagrange equations of the second kind may be found in (Landau and Lifschitz (1976)).

2.8   The Lagrange Equations with Damping

In the analysis that leads to the Lagrange equations it was assumed that the forces
can be obtained from a potential, implying that the system is conservative. Damping
is therefore not taken account of. Following an idea by *Bateman* in 1931, damping
can be included in a very elegant way (Bateman (1931)). The harmonic oscillator is
used as an example.

Bateman uses two systems in which one plays the role of a mirror to the other.
Thus when one of the systems experiences energy loss through damping, the other
system gains that energy by absorbing the same amount of energy. In this way the
total energy of the system is still conserved. To do the analysis, which is here limited
to one dimension, the coordinate of one system is called x and the coordinate of
the mirror is called $\bar{x}$; these are the canonical coordinates. Bateman put forward the
following Lagrangian:

$$\pounds(x, \bar{x}, \dot{x}, \dot{\bar{x}}) = \dot{x}\dot{\bar{x}} - \Omega^2 x\bar{x} + \gamma(x\dot{\bar{x}} - \bar{x}\dot{x}) \qquad (2.62)$$

Here the mass m has been assumed to be equal to unity. The generalised momenta
are

$$p = \frac{\partial \pounds}{\partial \dot{x}} = \dot{\bar{x}} - \gamma\bar{x}; \quad \bar{p} = \frac{\partial \pounds}{\partial \dot{\bar{x}}} = \dot{x} + \gamma x \qquad (2.63)$$

And the Lagrange equations become

$$\ddot{x} + 2\gamma\dot{x} + \Omega^2 x = 0; \quad \ddot{\bar{x}} - 2\gamma\dot{\bar{x}} + \Omega^2\bar{x} = 0 \qquad (2.64)$$

It is clear that the second equation is the time reversal of the first and the damping
from the first is absorbed by the negative damping of the second. Each equation
separately describes a damped harmonic oscillator. Energy conservation is illustrated
by multiplying the first with $\dot{x}$ and the second with $\dot{\bar{x}}$. It follows that

$$\frac{d}{dt}\left(\dot{x}^2 + \Omega^2 x^2 + \dot{\bar{x}}^2 + \Omega^2\bar{x}^2\right) = -2\gamma(\dot{x}^2 - \dot{\bar{x}}^2) \qquad (2.65)$$

which is zero if $\dot{x}^2 = \dot{\bar{x}}^2$. The Bateman approach stands as a curiosity in mechanics.
Attempts to extend the idea into other branches of physics, such as in quantum
mechanics where it would be a very advantageous application, have so far been
unsuccessful.

References

Bateman H (1931) On dissipative systems and related variational principles. Phys Rev 38:815–820
Kibble TWB (1985) Classical mechanics, 3rd edn. Longman Scientific and Technical, Harlow
Lagrange JL (1788) (Vol I), (1789) (Vol II) Méchanique Analytique. Courcier, Paris
Landau LD, Lifschitz EM (1976) Mechanics (3rd ed), Course of theoretical physics, vol 1. Pergamon, Oxford

Chapter 3
Two-Body Problem and the Solar System

Abstract The two-body problem is an important application, which is tackled in this chapter using Lagrange's equations with particular emphasis on the case of a central force field. The solution for the gravitational interaction yields Kepler's laws and the orbits of planetary motion. The relation between orbital elements is discussed. A short introduction to the solar system is given as this is obviously a very relevant case; some short historical notes are provided. The escape velocity is treated; the consequences of this for the presence of gases on planets and their moons are analysed. The perturbation of the planetary orbits due to the presence of nearby other planets is calculated. A brief analysis of Lagrange points and their stability is presented.

3.1 Two Bodies in a Central Force Field

Two bodies have masses m_1 and m_2. They have coordinates $\mathbf{r}_1 = (x_1, y_1, z_1)$ and $\mathbf{r}_2 = (x_2, y_2, z_2)$. The interaction is a function of the distance between the two only. This is an extremely important application in physics, for example, to describe the motion of the planets around the Sun. It is convenient to rephrase the problem in terms of centre of mass and relative coordinates. The centre of mass is located at

$$\mathbf{R} = \frac{m_1 \mathbf{r}_1 + m_2 \mathbf{r}_2}{m_1 + m_2} \tag{3.1}$$

The relative coordinate is $\mathbf{r} = \mathbf{r}_1 - \mathbf{r}_2$, so that

$$\mathbf{r}_1 = \mathbf{R} + \frac{m_2}{m_1 + m_2}\mathbf{r} \quad \mathbf{r}_2 = \mathbf{R} - \frac{m_1}{m_1 + m_2}\mathbf{r} \tag{3.2}$$

To describe two bodies in a central force field the potential depends on the magnitude of $\mathbf{r}$ only. It is then expedient to choose polar coordinates for the relative motion, while the motion of the centre of gravity can just remain in Cartesian coordinates. The kinetic energy is

Fig. 3.1 Sketch of the
two-body problem with polar
coordinates; $m_1 \gg m_2$

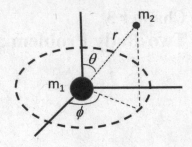

$$T = \frac{1}{2}(m_1 \dot{r}_1^2 + m_2 \dot{r}_2^2) = \frac{1}{2}(m_1 + m_2)\dot{\mathbf{R}}^2 + \frac{1}{2}\frac{m_1 m_2}{m_1 + m_2}\dot{\mathbf{r}}^2 \qquad (3.3)$$

The term $m_1 m_2 / (m_1 + m_2)$ is so central to the problem that it is given its own symbol μ; it is called the *reduced mass*. Polar coordinates are introduced, see Fig. 3.1. The term $\dot{\mathbf{r}}^2$ in polar coordinates is $\dot{r}^2 + r^2\dot{\theta}^2 + r^2\dot{\phi}^2 \sin^2\theta$.

The potential energy is a function of r only. Therefore the centre of gravity coordinates are cyclic and the associated momenta are conserved

$$(m_1 + m_2)\dot{\mathbf{R}} = \text{constant} \qquad (3.4)$$

In this way the centre of gravity motion is taken care of and need not be analysed further. It simply obeys Newton's first law. The Lagrangian for the relative motion is

$$\pounds = \frac{1}{2}\mu \left(\dot{r}^2 + r^2\dot{\theta}^2 + r^2\dot{\phi}^2 \sin^2\theta\right) - V(r) \qquad (3.5)$$

It is immediately clear that the angle ϕ is a cyclic coordinate and therefore the associated (angular) momentum is conserved; this quantity shall be called L

$$\frac{\partial \pounds}{\partial \dot{\phi}} = \mu r^2 \dot{\phi} \sin^2\theta = L \qquad (3.6)$$

The kinetic energy is homogeneously quadratic in the velocities, which makes the total energy E a conserved quantity

$$E = \frac{1}{2}\mu(\dot{r}^2 + r^2\dot{\theta}^2 + r^2\dot{\phi}^2 \sin^2\theta) + V(r) \qquad (3.7)$$

The rate of change of θ is best investigated in a special choice of the coordinate frame. Choose it in such a way that at some point in time t_0, when $\dot{\theta} = 0$ the orbit goes through $\theta = \pi/2$. The equation of motion associated with θ is

$$\frac{d}{dt}\left(\mu r^2 \dot{\theta}\right) - \mu r^2 \dot{\phi}^2 \sin\theta \cos\theta = 0; \quad \frac{d}{dt}\left(\mu r^2 \dot{\theta}\right) = 2\mu r\dot{r}\dot{\theta} + \mu r^2 \ddot{\theta} \qquad (3.8)$$

Now, at $t = t_0$ $\dot{\theta} = 0$, $\theta = \pi/2$ and therefore at this point $\ddot{\theta}$ vanishes. As a result, a time increment later both θ and $\dot{\theta}$ are both still zero. And so on to the next increment, which again results in an unchanged θ. All in all, the motion will take place in the plane defined by $\theta = \pi/2$. The whole problem is independent of θ, which leads to a tremendous simplification.

Summarising,

$$\mathcal{L} = \frac{1}{2}\mu(\dot{r}^2 + r^2\dot{\phi}^2) - V(r); \quad L = \mu r^2\dot{\phi}; \quad E = \frac{1}{2}\mu(\dot{r}^2 + r^2\dot{\phi}^2) + V(r) \quad (3.9)$$

Because θ is constant the orbit of the two bodies around each other is flat; it takes place in a plane. This result does not come as a surprise, as in a central potential the z-component of the angular momentum is the only one that is relevant, which means that the motion takes place in two dimensions. So, $L_z = L$.

The shape of the orbit can be established by eliminating the time, that is, $dt = (\mu r^2/L)^{-1}d\phi$ and hence $\dot{r} = L/(\mu r^2)(dr/d\phi)$. The energy integral is then

$$E = \frac{1}{2}\mu\left[\left(\frac{dr}{d\phi}\frac{L}{\mu r^2}\right)^2 + \frac{L^2}{\mu^2 r^2}\right] - \frac{\alpha}{r} \quad (3.10)$$

This equation prescribes a relationship between r and ϕ; it has a solution of the form

$$\frac{p}{r} = 1 + e\cos(\phi - \phi_0); \quad p = \frac{L^2}{\mu\alpha}; \quad e = \sqrt{1 + \frac{2EL^2}{\mu\alpha^2}} \quad (3.11)$$

For $-1 < e < 1$ it represents an ellipse, the centre of the coordinate frame residing in one of the focal points; e is called the *eccentricity* while $2p$ goes under the interesting name of *latus rectum*; p is the semilatus rectum, which is the distance between the focal point and the orbit, perpendicular to the long axis. The elliptical properties are illustrated in Fig. 3.2. The turning points of the orbit are found from $dr/d\phi = 0$. When the orbit describes a planet around the Sun, the point nearest the origin is called the *perihelion*—r_{per}—the furthest point is the *aphelion*, r_{ap}. For an orbit around the Earth, these points are termed *apigee* and *perigee*. (These terms are derived from the Greek words for Sun—$\eta\lambda\iota o\varsigma$—and Earth—$\gamma\eta$.) The long and the short axes of the ellipse are also quite easily found. So, to characterise the ellipse there is a choice of four sets of constants

1. The mechanical constants E and L.
2. The astronomical distances r_{ap} and r_{per}.
3. The geometrical constants p and e for the focal point as the origin.
4. The long and short axes of the ellipse a and b; the geometrical constants taking the centre of the ellipse as the origin.

The connections between all these are as follows

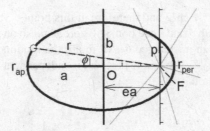

Fig. 3.2 Properties of the elliptical orbit

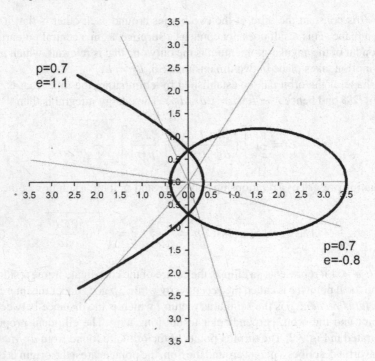

Fig. 3.3 Illustration of orbits

$$a = \frac{p}{1-e^2} = \frac{\alpha}{2|E|}; \quad b = \frac{p}{\sqrt{1-e^2}} = \frac{L}{\sqrt{2\mu|E|}} \qquad (3.12)$$

$$r_{ap} = \frac{p}{1+e} = a(1-e); \quad r_{per} = \frac{p}{1-e} = a(1-e) \qquad (3.13)$$

The energy E has been set in absolute values, because the condition on the eccentricity $-1 < e < 1$ can only be met if $E < 0$. If $E > 0$, the eccentricity is $|e| > 1$ and in that case the orbit is hyperbolic, which represents an object that enters from infinity, zooms around the origin and then disappears again into infinity (Fig.3.3).

For an elliptical orbit the period can be calculated. An infinitesimal area 'swept out' (that is, covered) over an orbital change $d\phi$ is $df = \frac{1}{2}r(rd\phi)$. Thus the rate of change of swept-out area is $\dot{f} = \frac{1}{2}r^2\dot{\phi}$. This is just proportional to the angular momentum L, so $L = 2\mu\dot{f}$. The area swept out in one whole revolution as ϕ goes through the interval $0 - 2\pi$ is the area of the ellipse πab. It is also equal to

$$\int_0^T \dot{f}dt = \frac{LT}{2\mu} \tag{3.14}$$

It follows that

$$T = \frac{2\pi\mu ab}{L} = 2\pi a^{3/2}\sqrt{\frac{\mu}{\alpha}} \tag{3.15}$$

3.2 The Solar System

Newton (1642–1726) published the *Philosophiae Naturalis Principia Mathematica* in 1687 (Newton 1687). 'Standing on the shoulders of giants' he derived *Kepler's laws* of planetary motion. Kepler published his work between 1609 and 1619; it was entirely based on observation (see *The Sleepwalkers* by Koestler 1986). Kepler's laws are

- The planets follow plane elliptical orbits with the Sun in the focal point.
- In equal times equal areas of the orbit are swept out.
- The relationship between the long axis of the ellipse and the period of the planet is the same for all planets: $T^2 \propto a^3$.

Newton managed to prove these relations, broadly speaking in the manner described above (though Lagrangian mechanics was not yet known, Lagrange 1736–1813). In order to prove Kepler's laws Newton had to introduce the following elements:

- Invent differentiation (a 'fluxion' calculus).
- Postulate Newton's laws.
- Hypothesise the interaction.

No sinecure, but in one brilliant move Newton solved the enormous planetary data set with the solution outlined in the previous section. It works because the Sun is so much heavier than the planets and may therefore be regarded as 'standing still'. A table of the properties of the planets is given below (Table 3.1).

The heaviest planets are Jupiter and Saturn. Even for these their masses are less than a thousandth of the Sun's. Therefore, the assumption by Kepler that all planets are in orbits around the same focal point is 99.9 percent correct. It is not only the planets that obey these laws, asteroids and comets do too. The asteroids are planetoids, mostly found between Mars and Jupiter; their masses are typically less than 0.1 percent of Mercury's mass (the largest one, Ceres, has a mass of $0.9 \times 10^{21} kg$). The comets have

Table 3.1 Planetary data for the solar system

Body	Mass (kg)	Mean radius (m)	Period (s)	Mean orbital radius (m)
Mercury	$3.1E23$	$2.4E6$	$7.60E6$	$5.79E10$
Venus	$4.88E24$	$6.06E6$	$1.94E7$	$1.08E11$
Earth	$5.98E24$	$6.37E6$	$3.16E7$	$1.50E11$
Mars	$6.42E23$	$3.37E6$	$5.94E7$	$2.28E11$
Jupiter	$1.90E27$	$6.99E7$	$3.74E8$	$7.78E11$
Saturn	$5.68E26$	$5.85E7$	$9.35E8$	$1.43E12$
Uranus	$8.68E25$	$2.33E7$	$2.64E9$	$2.87E12$
Neptune	$1.03E26$	$2.21E7$	$5.22E9$	$4.50E12$
Pluto	$\sim 1E23$	$\sim 3E6$	$7.82E9$	$5.91E12$
Sun	**1.99E30**	**6.96E8**	n/a	n/a

very elongated elliptical orbits; many travel to the outer regions of the solar system, the Oort belt, and have periods of 100 years or more. The well-known Halley's comet, for example, only visits the Sun every 70–80 years and is a 'short-period' comet. It is only seen once in a human life.

Another class of celestial bodies that obey Kepler's laws are the moons. These are bodies that orbit the planets. 'Our' moon, the Moon, is exceptionally large: the ratio of the diameter of the Moon to the diameter of the Earth is some 0.27. The sizes of the other satellites in the solar system are a much smaller fraction of the planet that keeps them in their gravitational hold. Neither Mercury nor Venus are believed to possess moons. The Earth has 1, Mars 2, Jupiter (possibly more than) 17, Saturn (possibly more than) 18, Uranus (possibly more than) 20, Neptune (possibly more than) 8 and Pluto (possibly more than) 5. All these bodies obey Kepler's laws. The management of man-made satellites that orbit the Earth or other planets obviously also relies on these simple mechanical laws. (All data from Patrick Moore's marvellous book *The data book of Astronomy*, Moore 2000.)

What would happen if the interaction V did not behave exactly as $1/r$, but more generally as r^n? It has been seen in the previous section that the orbit is exactly an ellipse and that the ellipse stands still in space. In other words, the planets go through the same trajectory over and over again. Only two interactions have that property: $n = -1$ and $n = 2$ (the latter case is the harmonic oscillator, the former is Newton's gravitational interaction). Every other exponent gives an orbit that does not go through the same trajectory time and again. When that occurs an effect, called *perihelion precession*, takes place, which is demonstrated in Fig. 3.4.

The planets in the solar system do indeed display some perihelion precession. For Mercury it is 43 s of arc per century and for Earth it is 3.8 s of arc per century. While these numbers are small—it takes Mercury over 2 million years to get back to the original orbit—they can be measured. Note that Mercury has quite an elliptical orbit, which helps. The question is then whether this effect can also be explained from

Fig. 3.4 Illustration of precession

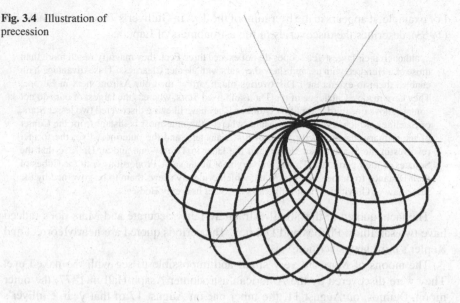

sources other than the one identified here, namely, $n \neq -1$. A number of explanations have been put forward:

- The Sun might have a systematic nonhomogeneous mass distribution. This is called the quadrupole moment of the Sun; there is no evidence that this might be a significant factor.
- Perturbation of the orbit of Mercury due to an undiscovered planet nearer the Sun. This has also been eliminated, because the type of extra motion that follows from such a presence is different than the perihelion precession.
- There is indeed something wrong with Newton's interaction, or with Newtonian mechanics.

While the latter reason is generally believed to be the true one, it is not easy to see directly how this should work. Einstein has shown that, indeed, a number of phenomena that have no Newtonian explanation can be understood if the character of gravity, space and time is slightly—but only ever so slightly—different than Newton had envisaged. Thus a footnote has been added to Newton's work.

When all orbits coincide to give a perfect ellipse, which is stationary in space, the solution is said to be degenerate. The degeneracy has an associated conserved quantity: the omega direction—the direction is the perihelion direction:

$$\Omega = \mathbf{v} \times \mathbf{L} + \alpha \frac{\mathbf{r}}{r} \tag{3.16}$$

Kepler's laws have been known for some time; he lived from 1571 to 1630. However, by the time Newton (1642–1727) came to prove Kepler's laws, the third one was pretty well forgotten. Due to Newton's work its importance was rediscovered.

For example, it appears in the literature of the day; in Gulliver's Travels (1726), Swift (1963), describes the discoveries of the astronomers of Laputa:

... although their largest Telescopes do not exceed three Feet, they magnify much more than those of a Hundred with us, and shew the Stars with greater Clearness. This Advantage hath enabled them to extend their Discoveries much farther than our Astronomers in Europe. They have made a Catalogue of ten Thousand fixed Stars, whereas the largest of ours do not contain above one third Part of that Number. They have likewise discovered two lesser Stars, or Satellites, which revolve about Mars; whereof the innermost is distant from the Center of the primary Planet exactly three of his Diameters, and the outermost five; the former revolves in the Space of ten Hours, and the latter in Twenty-one and an Half; so that the Squares of their periodical Times, are very near in the same Proportions with the Cubes of their Distance from the Center of Mars; which evidently shews them to be governed by the same Law of Gravitation, that influences the other heavenly Bodies.

The facts quoted in this small extract are fairly accurate and Mars does indeed have two satellites: Phobos and Deimos. The periods quoted are nearly correct and Kepler's third law is verified.

The moons of Mars are very small and impossible to see with the naked eye. They were discovered by the American astronomer Asaph Hall in 1877, the outer moon, Deimos, on August 11, the inner one on August 17 of that year. Gulliver's Travels was written—and this is baffling—in 1727! The other astonishing fact is that Phobos's period is less than the Martian day; out of the 30 or so satellites in the solar system that can be seen from Earth, there is one and only one for which this is true: Phobos—not so easy to predict. But then Gulliver's Travels is an amazing book; the cynical author who preferred horses to people, because he deemed the former honest (and who suggested that the hungry, child-rich Irish learn to eat babies) lets the professor from the Grand Academy of Lugado describe the memory matrix, a very twentieth-century device. He also suggests a form of language analysis that did not really get off the ground until the 1960s, because it requires a computer.

The Planetary Planes in the Solar System

The angle of inclination of the various orbits of the planets around the sun is listed in Table 3.2. It is interesting to note that the planes of the orbits of most planets in the solar system nearly coincide. The reason for this is of interest. A priori there is no reason why this should be so; the laws of mechanics suggest no strong mechanism that causes the orbits of neighbouring planets to align with one another. Yet it is generally believed that this is not a coincidence. The remarkable flatness of the solar system is possibly due to the manner in which it was formed in the first place. There are a number of hypotheses.

One option is that the solar system was formed due to an interaction between the Sun and another passing star. During the interaction hot gas was pulled out of the Sun and/or out of the other star and the remnants of this gas condensed to form the planets. A variant of this hypothesis is that the Sun is part of a double star system. The companion star would still be present and, according to a recent theory by Piet Hut, would currently be in the Oort belt. A long-term research programme to test this theory is being carried out. If the companion is found (it can be detected by checking its motion against the distant stars; the Sun's partner should move much more than all other stars), that would certainly be great news. So far, nothing has been found, but periodically the theory pops up in popular writings.

Table 3.2 Some of the orbital elements of the planets in the solar system

Body	Inclination (°)	Eccentricity	Escape velocity (km/s)
Mercury	7.0	0.205	4.1
Venus	3.4	0.007	10.0
Earth	0	0.017	11.2
Mars	1.8	0.093	5.0
Jupiter	1.3	0.048	59
Saturn	2.5	0.056	35
Uranus	0.8	0.047	22
Neptune	1.8	0.009	25
Pluto	17.2	0.250	?

Another possibility is that the Sun and the solar system were formed from the same spinning gas cloud, which was necessarily nonhomogeneous. The Sun contracted in the centre and the planets formed slowly by accretion of matter. A consequence of the approximate coincidence of all the planetary planes is that the planets always appear against the same background in the night sky. This background is a narrow belt called the *ecliptic* (the path of the Sun) and the 12 star constellations that appear at more or less equal angular spaces are called the signs of the Zodiac, namely, Aries, Taurus, Gemini, Cancer, Leo, Virgo, Libra, Scorpio, Sagittarius, Capricornus, Aquarius and Pisces.

The axis of the Earth makes an angle with the orbital plane and therefore, as the Earth spins, six signs appear to ascend to the North while six descend to the South. Currently, the ascending ones begin with Capricorn. The direction of the axis carries out a slow motion as if it describes the surface of a cone with a semi-top angle of 23.50^0; this motion is called *precession*. Thus the angle of the axis of the Earth's rotation with the ecliptic plane is not constant, but varies cyclically with a period of some 26,000 years. As a result the location of the Sun through the year moves up by about one sign every 2,100 years. For example, in early September the Sun was in Virgo around the birth of Christ, but is now in Libra. During the year the Sun traverses the signs in order: Aries, Taurus and Gemini are the spring signs; Cancer, Leo and Virgo are the summer signs; Libra, Scorpio and Sagittarius are the autumn signs and finally Capricorn, Aquarius and Pisces are the signs associated with the winter months.

Astrologers (magi) believe that the destiny of man is determined at birth by the position of the Sun in any of the Zodiacal constellations (though in some Latin countries, the relevant time point is supposed to be the conception). Strangely, the identity of the sign is evaluated as it would have been some 2000 years ago. The passage of the various planets through that constellation is viewed as an omen, or an oracle of some sort, which can be interpreted by the cognoscenti (Venus means love, Mars is trouble, etc.). As the motion of the planets is predictable from the historical record, careful observation proved worthwhile in ancient civilisations, and the life history of a famous person was usually prophesied, in suitably vague terms, by Royal Wise Men. The system was first conceived by the Babylonians who were quite reliable observers. Thanks to their meticulous method of recording we now have relatively accurate records of the positions of the planets, going back to *ca* 600 BC. The Chinese too recorded the motion of celestial bodies and extensive records have survived. Various ancient South American cultures have left astronomical data, but little is known about their astrological significance. Over the centuries the practice of astrology has come in for a good deal of criticism. À propos of the Chaldean practice the prophet Isaiah 47: 13–14 states (NIV)

[13] All the counsel you have received has only worn you out!

Let your astrologers come forward,

those stargazers who make predictions month by month,

let them save you from what is coming upon you.

[14] Surely they are like stubble; the fire will burn them up.

Similarly, Horace, 65BC-8BC, a Roman lyric poet, wrote the famous line *Nec Babylonios temptaris numeros* (Odes, Bk 1, xi, 2): Do not make trial of Babylonian calculations. A question is whether these ancients looked at the same sky compared to the location of the star signs in the twenty-first century. The location of the solar system in the Milky Way galaxy is such that in 6000 years it has rotated by some 0.001 degrees. While with modern instruments that might just be detectable, broadly speaking the background sky has hardly changed.

3.3 The Escape Velocity

In a simple one-dimensional analysis of the two-body problem it is investigated what happens when one of the bodies is very much heavier than the other. This is relevant to projectiles (rockets) being shot from a planet, or extraplanetary objects, such as asteroids, falling to a planet. Now $\mu \approx m$, the mass of the object and the energy is

$$E = -\frac{\alpha}{r} + \frac{1}{2}m\dot{r}^2 \qquad (3.17)$$

The case of interest is where $E = 0$; this case pertains to the problem in which the object has zero velocity at infinity. It implies that when, for example, a projectile is shot from Earth it will just be able to escape the Earth's gravity; if $E < 0$ the projectile will eventually fall back. When $E = 0$ the distance can be calculated as a function of time

$$r(t) = 2^{-1/3} \left(-3t\sqrt{\frac{\alpha}{m}} + r(0)^{3/2}\sqrt{2} \right)^{2/3} \qquad (3.18)$$

This relation is useful as an approximation for the distance as a function of time for comets, etc. approaching planets. The impact velocity is obtained directly from Eq. (3.17). If the target body has radius R and the comet is negligible in size compared to the target body, then the impact velocity is

$$\dot{r}|_R = -\sqrt{\frac{\alpha}{Rm}} \qquad (3.19)$$

Table 3.3 Average speeds of various gas molecules

Gas	Average velocity at 0° C (km/s)	Average velocity at −100° C (km/s)
Hydrogen	1.8	1.4
Helium	1.3	1.0
Methane	0.6	0.5
Carbon Dioxide	0.4	0.3
Water Vapour	0.6	0.5
Nitrogen	0.5	0.4
Oxygen	0.5	0.4

The minus sign is chosen to indicate that the body comes towards the target. The same relation with a plus sign is the *escape velocity*. That is the minimum speed required to let a body fly to infinity.

Other than the fuel-devouring minimum requirement for rocket speeds, the escape velocity is important in two other applications. The first is the interplanetary exchange of matter. It is possible for matter to fly from one planet to another if the fragments are given sufficient velocity. The energy for this event may be due to any explosive cause: meteorite impact, volcano eruption or the activity of a civilisation such as an atomic bomb exploding. So it is that small meteors from Mars are found on Earth. The second important application is the question of whether an atmosphere of some sort can exist on any planet. The velocity distribution of a gas is related to the temperature. If a substantial portion of the gas has a speed that exceeds the escape velocity, the atmosphere cannot exist, because over time any gas molecules will leak into space. The average molecular velocity of various gases at two relevant temperatures is tabulated in Table 3.3. The escape velocities for the various planets are listed in Table 3.2. As an added interesting point it is noted that the escape velocity for the Moon is 2.4 ms^{-1}. The Moon has no significant atmosphere as is clear from the comparison of the escape velocity with the average velocity of the atmospheric gases. Both the Earth and Venus have significant atmospheres, as do the outer planets.

3.4 Perturbation of Planetary Orbits

The discovery of Neptune happened because the orbit of Uranus displayed small irregularities. The French astronomer Leverrier calculated from these irregularities—wobbles in the orbit—that another planet should be present outside Uranus' orbit. He sent the result of his calculations to the German observer Galle in Berlin, who found Neptune that same evening. That was in 1846.

Pluto was discovered in 1930 by the American astronomer Clyde Tombaugh using a photographic technique. The calculations concerning the perturbation in the orbit

of Neptune that led to the discovery of Pluto were done by Percival Lowell, but he
died in 1916, 14 years before Pluto was discovered. Tombaugh died in 1996.

The problem that is studied here has the following features. Two planets are con-
sidered, the first with Cartesian coordinates $\mathbf{r}_1$, the second with $\mathbf{r}_2$. The interactions
in the problem are as follows:

Interaction of Planet 1 with the sun: $V_1 = -\alpha_1/|\mathbf{r}_1|$.

Interaction of Planet 2 with the sun: $V_2 = -\alpha_2/|\mathbf{r}_2|$.

Interaction of Planet 1 with Planet 2: $V_0 = -\alpha_0/|\mathbf{r}_1 - \mathbf{r}_2|$.

The kinetic energy is

$$T = \frac{1}{2}m_1\dot{\mathbf{r}}_1^2 + \frac{1}{2}m_2\dot{\mathbf{r}}_2^2 \tag{3.20}$$

To simplify the problem somewhat it will be assumed that the unperturbed orbits are
circular with radii R_1 and R_2. In addition, it is assumed that the orbits are in the same
plane. The perturbations in the radial motion are called u_1 and u_2; these are small in
magnitude compared to the radii. The other small parameter is α_0, as the masses of
the planets are much smaller than the mass of the Sun. The angular velocities will
be constants ω_1 and ω_2; the angular perturbations are denoted by ψ_1 and ψ_2. The
Lagrangian in these variables takes the form:

$$\pounds = \frac{1}{2}m_1\left(\dot{u}_1^2 + (R_1+u_1)^2\left(\omega_1+\dot{\psi}_1\right)^2\right) + \frac{1}{2}m_2\left(\dot{u}_2^2 + (R_2+u_2)^2\left(\omega_2+\dot{\psi}_2\right)^2\right) +$$
$$+\frac{\alpha_1}{R_1+u_1} + \frac{\alpha_2}{R_2+u_2} +$$
$$+\frac{\alpha_0}{\sqrt{(R_1+u_1)^2 + (R_2+u_2)^2 - 2(R_1+u_1)(R_2+u_2)\cos(\omega_1-\omega_2+\psi_1-\psi_2)}} \tag{3.21}$$

The Lagrange equation associated with u_1 is expanded up to first order in the perturbations

$$m_1\ddot{u}_1 - m_1R_1\left(\omega_1+\dot{\psi}_1\right)^2 + \frac{\alpha_1}{R_1^2} - m_1u_1\left(\omega_1+\dot{\psi}_1\right)^2 - 2\frac{\alpha_1 u_1}{R_1^3} +$$
$$+\frac{1}{2}\frac{\alpha_0(2R_1 - 2R_2\cos(\omega_1 t - \omega_2 t))}{\left(R_1^2 + R_2^2 - 2R_1R_2\cos(\omega_1 t - \omega_2 t)\right)^{3/2}} = 0 \tag{3.22}$$

Similarly, the Lagrange equation associated with ψ_1 in the same approximation is

$$m_1\frac{d}{dt}\left[(R_1+u_1)^2\left(\omega_1+\dot{\psi}_1\right)\right] - \frac{\alpha_0 R_1 R_2\sin(\omega_1 t - \omega_2 t)}{\left(R_1^2 + R_2^2 - 2R_1R_2\cos(\omega_1 t - \omega_2 t)\right)^{3/2}} = 0 \tag{3.23}$$

It follows immediately that

$$(R_1+u_1)^2\left(\omega_1+\dot{\psi}_1\right) = -\frac{\alpha_0}{(\omega_1-\omega_2)\sqrt{R_1^2 + R_2^2 - 2R_1R_2\cos(\omega_1 t - \omega_2 t)}} + \text{cnst} \tag{3.24}$$

The time-independent constant needs to be split into two parts. The first is associated with the mean motion and the second part with the perturbations, which are small. The former obviously equals $R_1^2\omega_1$. The latter, called c, needs to be derived from the initial conditions, which are determined from u_1 and $\dot{\psi}_1$ and this will remain as a constant in the equations.

The equations may be further developed by applying Kepler's third law, which is relevant to the unperturbed problem. Using $\alpha_1 = MGm_1$ and $\alpha_2 = MGm_2$ Kepler's third law reads

$$\omega_1^2 = \frac{MG}{R_1^3}; \quad \omega_2^2 = \frac{MG}{R_2^3}; \quad \frac{\omega_1}{\omega_2} = \frac{R_2^{3/2}}{R_1^{3/2}} \tag{3.25}$$

From (3.24) solve for $\dot{\psi}_1$

$$\dot{\psi}_1 = -\frac{(\omega_1 - \omega_2)\,(2R_1 m_1 \omega_1 u_1 + c)\,\sqrt{R_1^2 + R_2^2 - 2R_1 R_2 \cos(\omega_1 t - \omega_2 t)} + \alpha_0}{\sqrt{R_1^2 + R_2^2 - 2R_1 R_2 \cos(\omega_1 t - \omega_2)}\,(\omega_1 - \omega_2)\,R_1^2 m_1} \tag{3.26}$$

Substituting in (3.22)

$$m_1\ddot{u}_1 - m_1\omega_1^2 u_1 - 2\frac{\alpha_1 u_1}{R_1^3} + 2\frac{\omega_1(2R_1 m_1 \omega_1 u_1 + c)}{R_1} - F(t) = 0$$

$$\text{where } F(t) = \frac{2\alpha_0 \omega_1}{R_1\sqrt{R_1^2 + R_2^2 - 2R_1 R_2 \cos(\omega_1 t - \omega_2 t)}\,(\omega_1 - \omega_2)} +$$

$$+\alpha_0 \frac{R_1 - R_2 \cos(\omega_1 t - \omega_2 t)}{\left(R_1^2 + R_2^2 - 2R_1 R_2 \cos(\omega_1 t - \omega_2 t)\right)^{3/2}} \tag{3.27}$$

Using Kepler's third law the term that is proportional to u_1 can be rewritten in the simple form $m_1\omega_1^2 u_1$. The equation of motion is then completely analogous to an undamped, forced harmonic oscillator. The term that is proportional to α_0, the forcing function, is obviously periodic and proportional to $\cos(\omega_1 t - \omega_2 t)$. The silent assumption that has been made here is that at $t = 0$ the two unperturbed orbits of the planets are aligned. That is purely a matter of choosing the initial time and of no consequence for the form of the equations. If need be a phase can be introduced.

The forcing function $F(t)$ can be expanded in a Fourier series (see Sect. A.5) with basic period $\tau = 2\pi/(\omega_1 - \omega_2)$ and with coefficients

$$a_n = \frac{1}{\pi} \int_0^\tau F(t) \cos(n(\omega_1 - \omega_2)t)dt \tag{3.28}$$

By writing $R_r = R_2/R_1$ the function $F(t)$ may be scaled to α_0/R_1^2 and using Kepler's third law the scaled function then depends on R_r and t only. In the range $R_r > 1$ the first three scaled Fourier coefficients are plotted in Fig. 3.5); they have been calculated

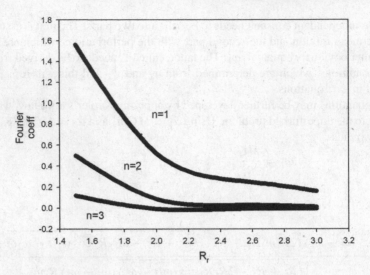

Fig. 3.5 Scaled Fourier components as a function of R_r

numerically. It is seen that in the relevant parameter range the coefficients diminish fast with increasing order n.

For each term in $F(t)$ of the form $a_n \cos(n(\omega_1 - \omega_2)t)$ there is a corresponding term in u_1 of magnitude

$$-\frac{a_n}{m_1(\omega_1^2 - n^2(\omega_1 - \omega_2)^2)} \tag{3.29}$$

The question then is what the ratio of the orbital periods are, especially in our solar system, and are there any that are very close to an integer that represents an appreciable Fourier coefficient. The only integers that are relevant in the solar system are 2 and 3 (the Jupiter/Mars ratio is 6.23) and the one that is very close to an integer is the ratio of the periods of Neptune to Uranus. A sample calculation will now be done on that system (there is the possible historical piquancy that the presence of Neptune was indeed predicted from the irregularities in the orbit of Uranus). The calculation is easily done with the aid of Table 3.1

$$\frac{u_U}{R_U} = -\frac{Gm_N}{R_U^3}a_2\frac{1}{\omega_U^2 - 4(\omega_U - \omega_N)^2} \tag{3.30}$$

where the subscripts U and N stand for Uranus and Neptune, respectively; G is the gravitational constant. Note that the factor $\omega_1^2/(\omega_1^2 - n^2(\omega_1 - \omega_2)^2)$ is approximately 44, so the upswing is tremendous. Using the numbers in the table the ratio u_U/R_U is some -0.4×10^{-3}. While that would seem to be a small value, the angular deflection of the orbit, as estimated from the value of $\dot{\psi}_U$, is some 8' (minutes of arc) in one orbit (that is approximately the time between the maximum and minimum in the oscillation with period $2\pi/(\omega_U - \omega_N)$); this is well within the observational capabilities of even a modest telescope.

The calculation for the perturbation of Uranus due to Neptune is exceptionally large because of the ratio of their orbital times, which resonates with the double frequency of the perturbation. This is a remarkable coincidence. Other perturbations in the solar system can also be calculated; these will be important when there are resonances, or when the masses are substantial (Jupiter and Saturn). Otherwise, the Sun dominates the gravitational interaction.

3.5 Lagrange Points

The previous section was concerned with the three-body problem, albeit under simplified assumptions. There is no simple analytical solution for the three-body problem, but there are other approximations that allow investigation revealing properties that are both insightful and useful. One such approximation is the investigation of so-called *Lagrange points*. In this approximation a two-body problem is used as a background configuration and the third body is permitted to navigate the potential field without disturbing the two-body problem very much. This is possible when the third body has a mass that is very much smaller than the masses involved in the two-body problem. The masses in the two-body problem are denoted by M and m and the mass of the much smaller third body is called m_0. The reduced mass of the masses M and m is $\mu = Mm/(M + m)$. A simple solution, sufficient for illustration purposes, is a flat circular orbit around the centre of mass. Let the orbit have a radius R then the circular frequency of the orbit is $\omega = \sqrt{mMG/(\mu R^3)}$. The origin of the rotating system is the centre of gravity. Let the location of the mass M be at $\mathbf{A}$ and the mass m at $\mathbf{a}$, then $\mathbf{A} = -m\mathbf{a}/M$. The ratio m/M will be called κ. A small mass m_0 located on the line from M to m has location vector $\mathbf{a}_0$, which may be written as $\mathbf{a}_0 = \lambda\mathbf{a}$. The assumption that it is on the line from M to m implies that it feels a centrifugal force $m_0\omega^2 r_0$, where r_0 is the coordinate from the origin (centre of mass of m and M) along the line. The force field working on m_0 consists of three contributions:

1. A potential due to M: $V_M = -GMm_0/|\mathbf{A} - \mathbf{a}_0|$, implying a force $-\partial V_M/\partial r_0$.
2. A potential due to m: $V_m = -Gmm_0/|\mathbf{a} - \mathbf{a}_0|$, implying a force $-\partial V_m/\partial r_0$.
3. The centrifugal force: $m_0\omega^2 r_0 = m_0mMG/(\mu|\mathbf{A} - \mathbf{a}|^3)$, as $R = |\mathbf{A} - \mathbf{a}|$.

The static force balance is

$$-GMm_0\frac{r_0 - A_x}{|\mathbf{A} - \mathbf{a}_0|^3} - Gmm_0\frac{r_0 - a_x}{|\mathbf{a} - \mathbf{a}_0|^3} + \frac{m_0mMGr_0}{\mu|\mathbf{A} - \mathbf{a}|^3} = 0 \tag{3.31}$$

In terms of the ratios κ and λ this becomes the requirement that the force $F(\lambda) = 0$

$$F(\lambda) = -\frac{\lambda + \kappa}{|-\kappa - \lambda|^3} - \kappa\frac{\lambda - 1}{|1 - \lambda|^3} + \frac{\lambda(1 + \kappa)}{|-\kappa - 1|^3} = 0 \tag{3.32}$$

The function is plotted for $\kappa = 0.1$ to illustrate the solution of the location of the points for which the force on the test mass vanishes. It is seen that there are three points on the line that goes through the masses m and M. The first point lies between the two masses and is called L_1; the second solution for $F(\lambda) = 0$ lies on the outside beyond the smaller mass and is called L_2; the third point, similarly, is found on the side of the heavy mass and is called L_3.

There are two more points for which the force balance vanishes and these two are located not on the line through m and M, but at an angle of $\pm\pi/3$ to it. The analysis is left to the reader, but they are called L_4 and L_5. The four points M, m, L_4 and L_5 form two equilateral triangles with the line m-M as the common base.

3.5.1 Stability of Lagrange Points

$F(\lambda)$ represents the force on the test mass, scaled to Gm_0Ma. From any of Lagrange points L_1, L_2 or L_3 a slight perturbation that moves its location in the positive direction by increasing λ would increase the force in the positive direction. This can be easily inferred from the graph in Fig. 3.6. Thus, the test mass is forced further in the positive direction. Similarly, a perturbation in the negative direction leads to a negative force and an associated motion in the minus direction. It is then concluded that these Lagrange points are not stable. A perturbation leads to an exponentially increasing function of the form $\exp(t/\tau)$. The value of the time constant τ can be derived from the steepness of the curve of the force as a function of λ and it will not

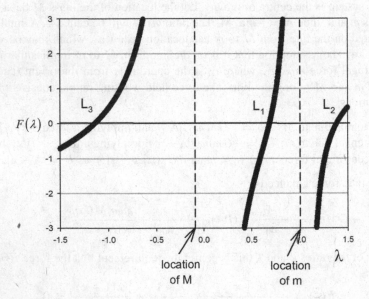

Fig. 3.6 The function $F(\lambda)$ for $\kappa = m/M = 0.1$

come as a big surprise that for L_1 and L_2 it is of the order of ω^{-1}. For L_3 the slope is less steep and therefore the time constant is rather larger and depends on the value of κ: $\tau \approx \omega^{-1}/\sqrt{8\kappa/3}$. For small mass ratios m/M.

In order to arrive at these constants the following method is used. Scale the potential and call it $\tilde{V}$; then expand locally around the equilibrium point, which is termed λ_0, as

$$\tilde{V}(\lambda) = \tilde{V}_0 + \frac{\partial \tilde{V}}{\partial \lambda}|_{\lambda_0}(\lambda - \lambda_0) + \frac{1}{2}\frac{\partial^2 \tilde{V}}{\partial \lambda^2}|_{\lambda_0}(\lambda - \lambda_0)^2 \tag{3.33}$$

In the equilibrium point $\partial \tilde{V}/\partial \lambda|_{\lambda_0} = 0$, so the dynamical equation becomes

$$\frac{d^2(\lambda - \lambda_0)}{dt^2} + \frac{\partial^2 \tilde{V}}{\partial \lambda^2}|_{\lambda_0}(\lambda - \lambda_0) \;\rightarrow\; \lambda - \lambda_0 \propto \exp(t/\tau)$$

$$\text{with } \tau = \left(-\frac{\partial^2 \tilde{V}}{\partial \lambda^2}|_{\lambda_0}\right)^{-1/2} \tag{3.34}$$

Now, for these unstable Lagrange points, the second derivative $\partial^2 \tilde{V}/\partial \lambda^2|_{\lambda_0}$ is negative and so there is a positive real value for τ.

To assess the stability of L_4 and L_5 the Coriolis force needs to be taken into account in addition to the centrifugal force. Surprisingly, when this is implemented in the analysis it leads to stability when the mass ratio M/m is greater than approximately 27, see Kibble (1985).

The Lagrange points play a significant role in space satellite management. When a spacecraft (which obviously has a much smaller mass than the planets or the Moon) is situated in a Lagrange point it feels no force, so it tends to stay there. For the unstable points L_1, L_2 and L_3 from time to time a corrective rocket blast has to be put into action, but for the stable points L_4 and L_5 even that is not necessary. In science fiction literature it is frequently assumed that L_3 is stable and that a satellite from an alien civilisation might reside there unnoticed. This point is defined by the Earth-Sun system. Such a satellite is called a *Trojan* and it could stay in that point for millions of years, surreptitiously observing us Earthlings. Unfortunately for the aliens L_3 is not a stable point.

Small space objects such as asteroids may congregate in the stable Lagrange points. Notably, for the Jupiter-Sun system, which has a period of some 11 years, so-called Trojan asteroids are present in L_4 and L_5. The mass ratio of the Sun to Jupiter is 1047. The Trojan asteroids are named after the heroes in the Trojan-Greek war, the Greek camp in L_4 and the Trojan camp in L_5. Other Sun-planet systems also host Trojan asteroids—often very small objects. These were discovered in the L_4 and L_5 points of the systems of the Sun with either Venus, Earth, Mars, Uranus or Neptune. The latter system plays host to many tens, the former quartet are populated by only one or two. The Sun-Jupiter system, however, is very richly populated with many thousands of bodies. An illustration of the distribution of asteroids near the orbit of Jupiter is shown in Fig. 3.7). For more details and lists of asteroids see Moore (2000).

Fig. 3.7 Distribution of
asteroids near the orbit of
Jupiter on 18-12-1997

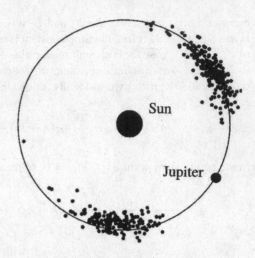

References

Kibble TWB (1985) Classical mechanics, 3rd edn. Longman Scientific and Technical, Harlow
Koestler A (1986) The sleepwalkers. Peregrine Books, Chicago
Moore P (2000) The data book of astronomy. Institute of Physics Publishing, Bristol
Newton I (1687) Philosophi Naturalis Principia Mathematica. The Royal Society (printer Joseph
    Streater), London
Newton I (1995) The principia (tr Motte A). Prometheus Books, New York
Swift J (1963) Gulliver's travels. Airmont Books, New York

Chapter 4
The Newtonian Gravitational Interaction

Abstract The Newtonian gravitational interaction is further investigated; refinements are discussed, especially in light of some twenty-first-century developments. Then Poisson's field equation is derived, which enables the calculation of a gravity field. A simple example of such a field is presented. The tidal force is introduced; it has the property that it pulls apart bodies that are near large gravitational objects. The Roche limit is mentioned. The disintegration of comets or asteroids is studied. A dynamic analysis is presented, which shows the importance of material damping on the disintegration process.

4.1 General Introductory Considerations

As noted in Chap. 1 the interaction due to gravity, as envisaged by Newton, is an inverse square law. This is what has been used to calculate the orbits of the planets around the Sun and the moons (and man-made satellites) around the planets. For these problems the bodies involved can be regarded as point masses. The potential energy associated with the gravitational interaction is like the Coulomb interaction in electrostatics with the masses playing the role of the charges

$$V(r) = -\frac{\alpha}{r}, \quad where \quad \alpha = GmM \tag{4.1}$$

The proportionality constant, called the gravitational constant, G has been measured to be $6.67259(85) \times 10^{-11} m^3 kg^{-1} s^{-2}$. Compared to other physical constants, the experimental physics community has struggled to measure G more accurately. The constant in Coulomb's law, for example, is known to a precision of ten decimal places. In part the uncertainty of G may be due to the very weak magnitude of the gravitational interaction, compared to other interactions. However, in a recent development it was shown that the value depends on the kind of experiment that is done to determine it.

In Fig. 4.1 the result of some recent experiments is depicted. There appears to be a substantial variation.

M. A. C. Koenders, *Constructing the Edifice of Mechanics*, Undergraduate Texts in Physics, https://doi.org/10.1007/978-3-031-34071-0_4

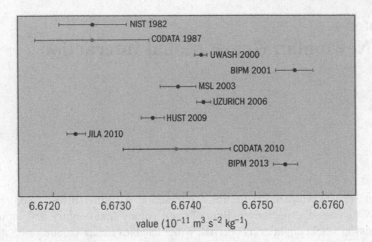

Fig. 4.1   Range of measurements of G (from (Cartwright 2014))

For the majority of mechanical features in the solar system the Newtonian interaction is a very accurate approximation. So, using this interaction the planetary orbits, the motion of satellites, *etc.* are all rather well understood. In the twentieth century a new theory of the gravitational interaction has been put forward by Einstein, which describes a number of second-order phenomena in the solar system. These include the perihelion precession of Mercury and the bending of light around the Sun, see Weinberg (1972). In our solar system, where mass densities are somewhat small, the effects of Einstein's theory are clearly almost negligible, but outside our solar system objects have been proven to exist with a phenomenal mass density and in the second half of the twentieth century and in the first couple of decades of the twenty-first century a host of effects have been discovered that can only be described using Einstein's theory. Among these is the direct measurement of gravitational waves, which requires excruciating experimental precision. These measurements, when perfected further, will open a new window on the Universe.

In the late 1980s it was briefly believed that there was a systematic error in the Eötvös experiment. It was argued that this error indicated that in addition to the long-range interaction there might be a short-range interaction as well: a weak $e^{(-r/\lambda)}$-type interaction. λ was believed to be of the order of a couple of hundred metres, too short to affect the extremely well-documented interplanetary motion and too long to notice it directly in the laboratory, where one might measure effects over a range of some ten metres. Now, many careful experiments later, the proposal of such a short-range force has been put on the back burner.

All in all, it can be said that there is still something deeply mysterious about the gravitational interaction.

4.2 Poisson's Equation

The question that must now be answered is: what happens when there are no point masses, but true, finite-sized bodies? For example, what is the gravitational potential inside the Sun? To answer this question a test mass m_t is introduced and a specific potential $U(\mathbf{x})$ of magnitude $U(\mathbf{x}) = V(\mathbf{x})/m_t$ is obtained by measuring the force on m_t at location $\mathbf{x}$. This force is due to the other masses in the field. A mass at location $\mathbf{a}$ results in a specific potential at $\mathbf{x}$ of magnitude

$$U(\mathbf{x}) = -G\frac{m(\mathbf{a})}{|\mathbf{x} - \mathbf{a}|}$$

(4.2)

N masses dotted around in the field result in a specific potential

$$U(\mathbf{x}) = -G\sum_{i=1}^{N}\frac{m_i(\mathbf{a}_i)}{|\mathbf{x} - \mathbf{a}_i|}$$

(4.3)

For a mass distribution with a position-dependent mass density $\rho(\mathbf{a})$ Eq. (4.3) is generalised by replacing the sum with an integral over the volume V

$$U(\mathbf{x}) = -G\int_V \frac{\rho(\mathbf{a})}{|\mathbf{x} - \mathbf{a}|}d^3a$$

(4.4)

Poisson's equation is the (differential) equation of which $U(\mathbf{x})$ is the solution. To obtain it, a technical detail needs to be addressed: when $\mathbf{x}$ approaches $\mathbf{a}$ the integral is ill-defined. To circumvent that problem add a small parameter ϵ to the distance function, that is,

$$|\mathbf{x} - \mathbf{a}| \rightarrow \sqrt{(x_1 - a_1)^2 + (x_2 - a_2)^2 + (x_3 - a_3)^2 + \epsilon}$$

(4.5)

Now differentiate Eq. (4.4) twice with respect to the coordinate x_j and sum over j

$$\frac{\partial^2 U(\mathbf{x})}{\partial x_j^2} = -3G\int_V d^3a\rho(\mathbf{a})\left[\frac{(x_j - a_j)^2}{|\mathbf{x} - \mathbf{a}|^5} - \frac{1}{|\mathbf{x} - \mathbf{a}|^3}\right]$$

(4.6)

With the use of (4.5) it is seen that this equation is equivalent to

$$\frac{\partial^2 U(\mathbf{x})}{\partial x_j^2} = -3G\int_V d^3a\rho(\mathbf{a})\frac{-\epsilon}{|\mathbf{x} - \mathbf{a}|^5} =$$

$$= -3G\int_V d^3a[\rho(\mathbf{a}) - \rho(\mathbf{x})]\frac{-\epsilon}{|\mathbf{x} - \mathbf{a}|^5} + 3G\rho(\mathbf{x})\int_V d^3a\frac{\epsilon}{|\mathbf{x} - \mathbf{a}|^5}$$

(4.7)

In the limit $\epsilon \to 0$ the first term vanishes. To evaluate the second term make the substitution $\mathbf{a} = \mathbf{x} + \mathbf{s}$. Then

$$3G\rho(\mathbf{x}) \int_V d^3a \frac{\epsilon}{|\mathbf{x}-\mathbf{a}|^5} = 3G\rho(\mathbf{x}) \int_V d^3a \frac{\epsilon}{|s|^5} = 12\pi G\rho(\mathbf{x}) \int_0^\infty ds\, s^2 \frac{\epsilon}{|s|^5} =$$

$$= 12\pi G\rho(\mathbf{x}) \int_0^\infty ds\, s^2 \frac{\epsilon}{|s^2+\epsilon|^{5/2}} = 12\pi G\rho(\mathbf{x}) \int_0^\infty ds \frac{s^2}{\epsilon^{5/2}} \frac{\epsilon}{|\frac{s^2}{\epsilon}+1|^{5/2}} =$$

$$= 12\pi G\rho(\mathbf{x}) \int_0^\infty dt \frac{t^2}{|t^2+1|^{5/2}} = 4\pi G\rho(\mathbf{x}) \tag{4.8}$$

Thus is established *Poisson's equation*, which relates the mass density to the specific potential field

$$\Delta U(\mathbf{x}) = 4\pi G\rho(\mathbf{x}) \tag{4.9}$$

A similar equation exists in the area of electrostatics Jackson (1962). The electric field strength $\mathbf{E}$ associated with an electric charge density ρ satisfies the conditions

$$\nabla \cdot \mathbf{E} = 4\pi\rho; \quad \nabla \times \mathbf{E} = 0 \tag{4.10}$$

The curl condition ensures that there is an electric potential Φ, such that $\mathbf{E} = -\nabla\Phi$, so that again Poisson's equation is obtained

$$\nabla^2 \Phi = -4\pi\rho \tag{4.11}$$

In the absence of charges Poisson's equation becomes *Laplace's equation*

$$\nabla^2 \Phi = 0 \tag{4.12}$$

So, there is a great similarity between electrostatics and classical static gravitational theory, which goes back to the inverse square law, which has the same form for point charges: Newton's gravitational law and Coulomb's electrostatic law.

4.2.1   U(r) for a Homogeneous Spherical Body

Let the mass density for a body with radius R be constant ρ_0. Symmetry implies that there is no angular dependence, so Poisson's equation depends on the radial coordinate only; call this one r, then

$$\frac{1}{r^2}\frac{\partial}{\partial r}\left(r^2\frac{\partial U}{\partial r}\right) = 4\pi G\rho(r) \tag{4.13}$$

Here the mass density function is $\rho(r') = \rho_0$ if $r' \le R$ and $\rho(r') =$ if $r' > R$. Integrating gives

$$\frac{\partial U}{\partial r} = \frac{4\pi}{3} G\rho_0 r, \quad (r \le R); \quad \frac{\partial U}{\partial r} = \frac{4\pi}{3} G\rho_0 \frac{R^3}{r^2}, \quad (r > R) \tag{4.14}$$

The total mass of the body is $M = \frac{4\pi}{3}\rho_0 R^3$, so for $r > R$ the limit of point masses is again applicable.

4.3 Tidal Forces

Everybody on Earth is continually aware of the gravitational interaction. Things drop and fall all the time; it dominates the existence of everything that lives on the planet. Although compared to other interactions known to physics it is definitely a weak force; the cumulative effect of all the masses in the planet makes its manifestation ubiquitous. This is unlike the electrical force, which—though very much stronger— is hardly manifested on a macroscopic scale, because there is no accumulation, as positive and negative charges are present in virtually equal amounts in atoms and the net effect is small. The only time when the electrical force is making its strong presence felt is when there is a substantial separation of charges, such as it occurs during thunderstorms or when the charges are artificially manipulated in Man's harnessing of electrical power.

A feature of life on Earth is the occurrence of the *tides*. These are associated with the Moon and, again, extremely well known. Note that by planetary standards in the solar system the Earth's Moon is quite a large satellite. The tidal effect is easily explained. Due to the distance-dependence of the gravitational interaction, parts of the participating bodies that are closer together feel a larger attraction than those that are further apart. In this way a differential potential is created, which gives rise to a repelling force between the parts. Thus a body that is near a planet will feel a force that pulls it apart. This force may to a greater or lesser extent be balanced by the gravitational self-attraction.

The analysis is easily done for a planetoid sphere of radius R with uniform mass density ρ_0 in the vicinity of a planet of density ρ_t of radius R_t (the subscript t stands for 'target'). First the self-attraction is calculated. The specific potential has already been calculated. From Eq. (4.14) it follows that for $r < R$

$$U(r) = \frac{2\pi}{3} G\rho_0 r^2 = \frac{2\pi}{3} G\rho_0 \left(x^2 + y^2 + z^2\right) \tag{4.15}$$

The force per unit volume along the z-axis through the centre of the sphere is therefore

$$f(z) = -\frac{4\pi}{3} G\rho_0 z \tag{4.16}$$

The force on a unit mass in the planetoid due to the attraction of the nearby planet at distance R_c is

$$f_t = -\frac{4\pi}{3}\rho_t \frac{GR_t^3}{(R_c + z)^2} \tag{4.17}$$

It may be assumed that $|z| \ll R_c$, so a Taylor expansion is appropriate

$$f_t = -\frac{4\pi}{3}\rho_t \frac{GR_t^3}{R_c^2} + \frac{8\pi}{3}\rho_t \frac{GR_t^3 z}{R_c^3} \tag{4.18}$$

The first term is the force per unit mass on the planetoid centre of gravity; it pulls the planetoid towards the planet. The second term is the internal force and is combined with the self-attraction. The effective force per unit mass is thus

$$-\frac{4\pi}{3}G\rho_0 z + \frac{8\pi}{3}\rho_t \frac{GR_t^3 z}{R_c^3} \tag{4.19}$$

While this is negative the planetoid is pulled together, but when the expression becomes positive, ignoring any material cohesion, the planetoid can fall apart. The critical distance at which the effective force vanishes is known as the *Roche limit*

$$R_c = R_t \left(\frac{2\rho_t}{\rho_0}\right)^{1/3} \tag{4.20}$$

The Roche limit gives a first-order approximation of the stability of satellites around planets, etc. It is easily checked that the Moon, for example, is outside the Roche limit and will therefore not break up under the tidal forces exerted by the Earth. When the Schumacher-Levy asteroid in 1992 came within Jupiter's Roche limit it broke in three pieces (it hit the planet in 1994).

4.3.1 The Dynamics of Planetoid Fracture Due to Tidal Forces

Plenty of objects that approach a planet do not break up inside the Roche limit. There is ample evidence of such objects, even in recent times, that have hit (or have come close to) Earth, for example. The 1904 explosion of a comet or asteroid over Tunguska in Siberia, the Chelyabinsk meteor (a superbolide) that entered Earth's atmosphere over Russia on 15 February 2013 and the 2020 meteorite fireball over Japan are some well-documented instances. The (in)famous asteroid impact 66 million years ago that caused the Chicxulub crater on the Yucatan Peninsula in Mexico is another Earth-related example that acquired popular notoriety.

The idea that the gravitational self-attraction describes the physics may not be sufficient. Asteroids and meteors are often very solid bodies. Their encounter with a tidal force is frequently betrayed by their weird shapes, which shows that they have been deformed into a kind of irregular dumbbell profile. Here, it will be attempted to model the approach of a planetoid that retains its solid mechanical properties. Also, Roche's concept of a static approach to the problem is obviously wrong when a planetoid is hurled towards the target body. The question then is how the stability of such objects can be described. The approach taken is to study how a small initial perturbation develops as the approach takes place.

The planetoid is supposed to consist of two parts, each having a mass m. These are located at distances z_1 and z_2 from the centre of the target. The potential energy associated with the gravitational interaction of the two parts is

$$V = -\frac{\alpha}{z_1} - \frac{\alpha}{z_2}, \tag{4.21}$$

where $\alpha = GM_t m$; G is the gravitational constant, M_t is the mass of the target. Centre of gravity and relative coordinates may be introduced: $R_c = \frac{1}{2}(z_1 + z_2)$ and $z = \frac{1}{2}(z_1 - z_2)$.

Now, it may be assumed that $|z| << |R_c|$, so that in a Taylor expansion

$$V = -2\alpha \left(\frac{1}{R_c} + \frac{z^2}{R_c^3} \right) \tag{4.22}$$

4.3.1.1 Centre of Gravity Motion

The centre of gravity motion (again assuming $|z| << |R_c|$ and also that $\dot{R}_c(\infty) = 0$) is found from

$$\ddot{R}_c = -\frac{GM_t}{R_c^2} \quad \rightarrow \quad R_c(t) = \left(R_0^{3/2} - 3t\sqrt{\frac{\alpha}{2m}} \right)^{2/3} \tag{4.23}$$

Here R_0 is a reference distance. Instead of the time t an alternative time scale T may be introduced

$$T = \frac{1}{3} R_c^{3/2} \sqrt{\frac{2m}{\alpha}}, \quad \rightarrow \quad t = \frac{1}{3}\sqrt{\frac{2m}{\alpha}} R_0^{3/2} - T \tag{4.24}$$

4.3.1.2 Relative Motion

As yet no term for the self-attraction of the planetoid has been introduced. Here, a very simple form is chosen: a spring-type interaction with potential $V = \frac{1}{2}kz^2$. For most geological materials the value of k is such that the gravitational self-attraction is negligible. In addition a damping term is added with a force-velocity dependence $\gamma\dot{z}$. The equation of motion for the relative motion then becomes

$$2m\ddot{z} + \left(k - \frac{4\alpha}{R_c(t)^3}\right)z + \gamma\dot{z} = 0 \qquad (4.25)$$

Instead of the time t the alternative time scale T is used and then $R_c(t)$ is also expressed in this variable. Differentiation with respect to T is denoted by a $'$.

$$2mz'' + \left(k - \frac{8}{9}\frac{m}{T^2}\right)z - \gamma z' = 0 \qquad (4.26)$$

Something akin to the Roche limit can be seen. The point, when $T = T_c$ and $R_c = R_{cc}$, at which the term proportional to z is zero, when the 'effective spring constant' changes sign:

$$T_c^2 = \frac{8m}{9k}, \ or \ R_{cc}^3 = \frac{4\alpha}{k} = \frac{4M_t mG}{k} \qquad (4.27)$$

The solution to the linear differential Eq. (4.26) depends on the severity of the damping.

$$\text{When } \frac{\gamma^2}{16\,m^2} < \frac{4}{9T_c^2}$$

$$z(T) = e^{\frac{\gamma T}{4m}}\sqrt{\frac{T}{T_c}}\left[AJ_{\frac{5}{6}}\left(T\sqrt{-\frac{\gamma^2}{16\,m^2} + \frac{4}{9T_c^2}}\right) + BY_{\frac{5}{6}}\left(T\sqrt{-\frac{\gamma^2}{16\,m^2} + \frac{4}{9T_c^2}}\right)\right]$$

$$(4.28)$$

$$\text{When } \frac{\gamma^2}{16\,m^2} > \frac{4}{9T_c^2}$$

$$z(T) = e^{\frac{\gamma T}{4m}}\sqrt{\frac{T}{T_c}}\left[AI_{\frac{5}{6}}\left(T\sqrt{\frac{\gamma^2}{16\,m^2} - \frac{4}{9T_c^2}}\right) + BK_{\frac{5}{6}}\left(T\sqrt{\frac{\gamma^2}{16\,m^2} - \frac{4}{9T_c^2}}\right)\right]$$

$$(4.29)$$

Here $J_{\frac{5}{6}}$ and $Y_{\frac{5}{6}}$ are Bessel functions of order $5/6$ and $I_{\frac{5}{6}}$ and $K_{\frac{5}{6}}$ are modified Bessel functions of order $5/6$, see Sect. A.8.3, Abramowitz and Stegun (1972), Boas (1983). A and B are the initial conditions.

4.3.1.3 Sensitivity to a Small Perturbation

The solutions (4.28) and (4.29) can be used to study to small perturbations. The Bessel functions are, generally speaking, well known. However, in combination with the square root and the exponential a small study is appropriate. To begin with the damping is set to zero; in that case only (4.28) is relevant. The functions $\sqrt{y}J_{\frac{5}{6}}(y)$ and $\sqrt{y}Y_{\frac{5}{6}}(y)$ are plotted to reveal their character. When the damping is dominant, the relevant functions are in (4.29); they are of the form $e^y\sqrt{y}I_{\frac{5}{6}}(y)$ and $e^y\sqrt{y}K_{\frac{5}{6}}(y)$. These functions are shown in Fig. 4.2.

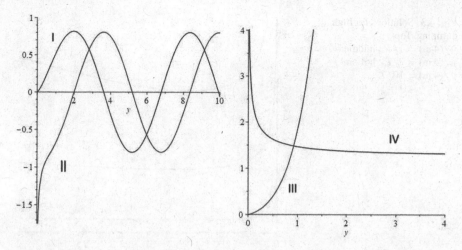

Fig. 4.2 Functions in solutions (4.28) and (4.29). I: $\sqrt{y}J_{\frac{5}{6}}(y)$, II: $\sqrt{y}Y_{\frac{5}{6}}(y)$; III: $e^{y}\sqrt{y}I_{\frac{5}{6}}(y)$; IV: $e^{y}\sqrt{y}K_{\frac{5}{6}}(y)$

The motion of the planetoid goes from $T \to \infty$ to $T \to 0$. A small fluctuation is seen to increase for the cases involving the functions $Y_{\frac{5}{6}}$ and $K_{\frac{5}{6}}$ as these show a blowing up as T decreases and the planetoid approaches the target. The relative motion described by solution (4.28) is oscillatory, while that associated with (4.29) is monotone in nature. When developing a narrative of the approach to the target, the material parameters do not have to be assumed to be constants; a solution can be picked up at any point by fitting a subsequent solution with alternative parameters by choosing new values for A and B. The actual quantitative value of the displacements is not really relevant; it is the *character* of the solutions that matter.

A plausible narrative is the following. When the planetoid is far away—y large in Figs. (4.2) and (4.3)—the damping is a small viscous value, so there are small oscillations in the relative motion: solution (4.28). The object is probably quite heterogeneous, so the relative motion will cause friction, which raises the temperature. Fragments of the object will then be expelled. An object that loses mass is subject to higher damping; k will also go down. This effect will then push the material parameters into the high damping régime: solution (4.29). In Fig. 4.3 the solution for the term that is proportional to B is plotted as a function of T/T_c. The relative motion will initially be diminished until, quite suddenly, when T/T_c has become smaller than approximately 0.5 the relative motion explodes. At that point the current linear equations obviously do not describe events very accurately any more. Other effects, such as the interaction of the planetoid with an atmosphere, have also not been represented. However, the sudden explosion well inside the traditional Roche limit has been observed on a number of occasions. In some cases no impact crater has been created, as the planetoid exploded into many fragments just above the surface of the target.

Fig. 4.3 Solutions for high
damping. Top:
$\gamma/(4\,\mathrm{m}) = 1/T_c$; middle:
$\gamma/(4\,\mathrm{m}) = 2/T_c$; bottom
$\gamma/(4\,\mathrm{m}) = 10/T_c$

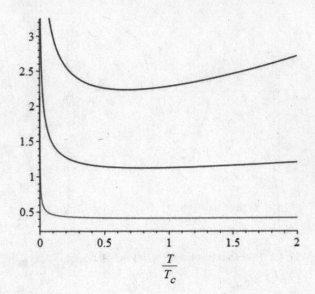

References

Abramowitz M, Stegun A (1972) Handbook of mathematical functions. Dover, New York
Boas ML (1983) Mathematical methods in the physical sciences. Wiley, New York
Cartwright J (2014) The lure of G. Physics World. 13 Feb. https://physicsworld.com/a/the-lure-of-
    g/
Jackson JD (1962) Classical electrodynamics. Wiley, New York
Weinberg S (1972) Gravitation and cosmology. Wiley, New York

Chapter 5
Mechanics Using Integral Principles

Abstract The theory of mechanics is further developed and phrased as a variational problem. Euler's calculus of variations is introduced and Hamilton's principal function is treated as an example. It is shown that the Lagrange equations (of the second kind) can be derived from a variational principle. Using generalised coordinates and generalised momenta the Hamiltonian and Hamilton's equations are obtained. Poisson brackets are used to derive the development of a mechanical variable as a function of time. Canonical transformations are discussed. The action integral is investigated and the Hamilton-Jacobi equations are derived as yet another alternative to obtaining the equations of motion. This piece of physics leads to a way of obtaining useful conserved quantities. Examples of the Hamilton-Jacobi equations are given and a new insight in the propagation of light is acquired, which leads to an understanding of the wave-particle duality.

5.1 Calculus of Variations

The theory of mechanics due to Newton and Lagrange was in its essential form known before 1800. In the nineteenth century Hamilton and Jacobi (independently from one another) developed schemes to integrate the equations of motion. These schemes are not generally used to solve practical problems in classical mechanics, however they have turned out to be very important for the development of physics, especially in quantum mechanics and relativity theory. To begin with the *calculus of variations* is introduced. A function $y(x)$ is to be determined on an interval x_1, x_2, such that the integral I is stationary

$$I = \int_{x_1}^{x_2} F\left(x, y, \frac{dy}{dx}\right) dx \qquad (5.1)$$

Here the function $F(x, y, y')$ is given and $y(x_1)$ and $y(x_2)$ have prescribed values. The idea is to obtain a *condition* for the function F for when I is stationary, that is, $\delta I = 0$. Finding the condition is achieved by adding an arbitrary function $\alpha\eta(x)$ to the function $y(x)$; $\eta(x)$ vanishes in x_1 and x_2. Expression (5.1) is then a function of

© The Author(s), under exclusive license to Springer Nature Switzerland AG 2023
M. A. C. Koenders, *Constructing the Edifice of Mechanics*, Undergraduate Texts
in Physics, https://doi.org/10.1007/978-3-031-34071-0_5

the parameter α. The condition $\delta I = 0$ is

$$\left(\frac{dI}{d\alpha}\right)_{\alpha=0} = 0 \qquad (5.2)$$

This implies that

$$\int_{x_1}^{x_2} \left[\frac{\partial F}{\partial y}\eta(x) + \frac{\partial F}{\partial y'}\eta'(x)\right] dx = 0 \qquad (5.3)$$

Partial integration of the second term leads to

$$\int_{x_1}^{x_2} \left[\frac{\partial F}{\partial y} - \frac{d}{dx}\left(\frac{\partial F}{\partial y'}\right)\right] \eta(x)dx = 0 \qquad (5.4)$$

As $\eta(x)$ is arbitrary the variation problem $\delta I = 0$ implies that the function $y(x)$ must satisfy *Euler's equation*

$$\frac{\partial F}{\partial y} - \frac{d}{dx}\left(\frac{\partial F}{\partial y'}\right) = 0 \qquad (5.5)$$

Extensions to the formalism are easily made.

5.1.1 Variation Problem with n Functions

When there are n functions of t rather than x, called $q_1(t), q_2(t), ..., q_n(t)$ then for each of these it holds that

$$\delta I = \delta \int_{t_1}^{t_2} F(t, q_k, \dot{q}_k)dt = 0 \rightarrow \frac{d}{dt}\left(\frac{\partial F}{\partial \dot{q}_k}\right) - \frac{\partial F}{\partial q_k} = 0 \ (k = 1..n) \qquad (5.6)$$

It is observed that the Lagrange equations of the second kind are equivalent to Euler's equations that pertain to the variation problem

$$\delta \int_{t_1}^{t_2} \pounds dt = 0 \qquad (5.7)$$

5.1.2 Extension to More Variables

When the functions depend on more than one variable the formalism also works. Suppose there are two variables, x and t. The integral boundaries are now the edge of an area A and the condition for stationarity is

$$\delta \int \int_{\partial A} F(t, x, y, y_t, y_x)dtdx = 0 \qquad (5.8)$$

Euler's equation then reads

$$\frac{\partial}{\partial t}\frac{\partial F}{\partial y_t} + \frac{\partial}{\partial x}\frac{\partial F}{\partial y_x} - \frac{\partial F}{\partial y} = 0 \tag{5.9}$$

An example is the equation for the vibrating string in one dimension. The mass density is ρ and the elasticity modulus E. The Lagrangian density is

$$\pounds = \frac{1}{2}\rho\left(\frac{\partial y}{\partial t}\right)^2 - \frac{1}{2}E\left(\frac{\partial y}{\partial x}\right)^2 \tag{5.10}$$

Leading to the wave equation

$$\rho\frac{\partial^2 y}{\partial t^2} - E\frac{\partial^2 y}{\partial x^2} = 0 \tag{5.11}$$

Further extensions involving higher derivatives can also be explored.

5.1.3 Extension to Problems with Imposed Conditions

Here the simple case is treated of one imposed condition and a function that depends on one variable x only. Extensions to more conditions and higher dimensional problems are easily obtained.

The condition reads $G(x) = 0$, the variational problem is extended to include a Lagrange multiplier λ. Create the functional $F(y_x, y, x) + \lambda G(y, x)$ and require

$$\delta\int_{x_1}^{x_2} (F(y_x, y, x) + \lambda G(y, x))\,dx = 0 \tag{5.12}$$

Euler's equation now takes the form

$$-\frac{d}{dx}\left(\frac{\partial F}{\partial y_x}\right) + \frac{\partial F}{\partial y} + \lambda\frac{\partial G}{\partial y} = 0 \tag{5.13}$$

5.2 Hamilton's Principle

The variational problem can be used to rephrase the equations of motion in mechanics. It was already established that the variational principle can be used to arrive at Lagrange's equations. The formalism is extended to include imposed conditions. Literature on this subject may be found in, for example, Kibble (1985), Landau and Lifshitz (1976).

Lagrange's equations of the first kind are

$$m_\nu \ddot{\mathbf{r}}_\nu - \mathbf{F}_\nu + \sum_{j=1}^p \lambda_j \nabla_\nu G_j = 0, \qquad G_j(\mathbf{r}_\nu, t) = 0 \tag{5.14}$$

For the case that the force $\mathbf{F}$ can be given by a potential the equation of motion becomes

$$\frac{d}{dt}\frac{\partial \pounds}{\partial \dot{x}_\nu} - \frac{\partial \pounds}{\partial x_\nu} + \sum_{j=1}^p \lambda_j \nabla_\nu G_j = 0 \tag{5.15}$$

and similar for y_ν and z_ν.

These same equations would have been followed from the Euler equation associated with the variational principle

$$\delta \int_{t_1}^{t_2} \pounds(\mathbf{r}_\nu, \dot{\mathbf{r}}_\nu, t) dt = 0, \qquad G_j(\mathbf{r}_\nu, t) = 0 \tag{5.16}$$

Therefore, the Euler equation associated with the variational principle is equivalent to Lagrange's equations.

When generalised coordinates q_k are employed no imposed conditions are necessary. Then the Euler equations are just Lagrange's equations of the second kind. This is known as *Hamilton's principle*. For a holonomic system that possesses a Lagrangian the Lagrange equations follow from the variational principle, a very important finding as the *choice* of generalised coordinate(s) does not appear to matter.

The integral can be interpreted as a function of the initial coordinates, end coordinates and the time difference

$$\int_{t_1}^{t_2} \pounds dt = W(q_k(t_2), q_k(t_1), t_2 - t_1) \tag{5.17}$$

Expressed in this way W is called *Hamilton's Principal Function*.

In physics there are a number of such integral principles that all have the dimension of an energy times the time, which is a so-called *action*. Such a principle is termed a *principle of least action*. The word 'least' may be slightly misleading, as it does not always concern a minimum; the variational principle requires the *stationarity* of the integral.

5.2.1 Optical Paths

In order to understand the principle of least action a little better, an analogous problem is investigated. In optics the *optical path length* is defined as the refractive index n times the length of the path that a light ray travels. An increment of optical path

Fig. 5.1 Sketch to illustrate
the refractive index problem

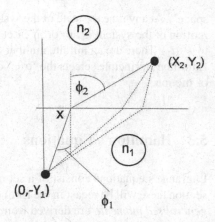

length is therefore nds. Now consider a situation in which light must travel from a
point in one medium with refractive index n_1 to another point in another medium
with refractive index n_2. Let the boundary of separation between the two media be
at $y = 0$. The point of departure has coordinates $(0, -Y_1)$ and the point of arrival is
at X_2, Y_2, see Fig. 5.1.

The question is: which path will the light ray take. In the figure various possible
paths are indicated as the light would appear to have an enormous variety of options.
However, it chooses one particular path and that is the one for which the optical path
length is minimal. In other words,

$$\delta \int_{P_1}^{P_2} nds = 0 \tag{5.18}$$

It is easy to see that this is equivalent to

$$\left(n_1\sqrt{x^2 + Y_1^2} + n_2\sqrt{(X_2 - x)^2 + Y_2^2} \right) \quad \text{minimal} \tag{5.19}$$

Differentiating with respect to x and requiring the result to be zero then leads to
Snell's Law of refractive indices

$$n_1 \sin \phi_1 = n_2 \sin \phi_2 \tag{5.20}$$

5.2.2 Geometrical Interpretation of the Action Principle

The geometrical interpretation of Hamilton's principle of least action is analogous
to the analysis of the paths of light rays in the previous section. Consider a Cartesian
space spanned by the generalised coordinates q_k (sometimes called the 'configuration

space'). At any time the state of the system is represented by a point. In this space the motion of the system is an 'orbit'. Let the system now move from one point $P_1(t_1)$ to $P_2(t_2)$. There are an infinite number of possible orbits between the two points, but Hamilton's principle selects the 'true' one, that is, the one that satisfies the equation of motion.

5.3 Hamilton's Equations

Lagrange's equations consist of a set of second-order differential equations. In this section these will be recast in the form of two sets of equations of the first order. The *generalised momenta* are derived from the Lagrangian

$$p_k = \frac{\partial £}{\partial \dot{q}_k} \tag{5.21}$$

From this $\dot{q}_k$ can in principle be solved and expressed in p_k. A new function is now created, called the *Hamiltonian*

$$H(p_k, q_k, t) = \sum_k p_k \dot{q}_k - £(q_k, \dot{q}_k, t) \tag{5.22}$$

If the potential energy V is independent of the velocities and the kinetic energy T is homogeneously quadratic in the velocities then the Hamiltonian H is equal to the total energy

$$H(p_k, q_k, t) = \sum_k p_k \dot{q}_k - £(q_k, \dot{q}_k, t) = \sum_k \frac{\partial T}{\partial \dot{q}_k} \dot{q}_k - £ = 2T - T + V = T + V$$
$$\tag{5.23}$$

Furthermore,

$$dH = \sum_k p_k d\dot{q}_k + \sum_k \dot{q}_k dp_k - \sum_k \frac{\partial £}{\partial q_k} dq_k - \sum_k \frac{\partial £}{\partial \dot{q}_k} d\dot{q}_k - \frac{\partial £}{\partial t} dt =$$
$$= \sum_k \dot{q}_k dp_k - \sum_k \frac{\partial £}{\partial q_k} dq_k - \frac{\partial £}{\partial t} dt \tag{5.24}$$

And so it follows

$$\dot{q}_k = \frac{\partial H}{\partial p_k}; \quad -\frac{\partial £}{\partial q_k} = \frac{\partial H}{\partial q_k}; \quad -\frac{\partial £}{\partial t} = \frac{\partial H}{\partial t} \tag{5.25}$$

It is seen that if the Lagrangian is independent of t, the Hamiltonian is also time-independent. Furthermore, if q_k does not appear in the Lagrangian (a cyclic coordinate) it does not appear in the Hamiltonian either.

Now, from Lagrange's equations

$$\dot{p}_k = \frac{\partial \pounds}{\partial q_k},$$ (5.26)

it follows that

$$\dot{q}_k = \frac{\partial H}{\partial p_k}; \quad \dot{p}_k = -\frac{\partial H}{\partial q_k}$$ (5.27)

These are called *Hamilton's equations* or the *canonical equations of motion*.

Conservation of energy is studied by taking the total derivative of the Hamiltonian with respect to time

$$\frac{dH}{dt} = \sum_k \left(\frac{\partial H}{\partial q_k} \dot{q}_k + \frac{\partial H}{\partial p_k} \dot{p}_k \right) + \frac{\partial H}{\partial t}$$ (5.28)

Substituting Hamilton's equations then yields

$$\frac{dH}{dt} = \frac{\partial H}{\partial t}$$ (5.29)

In other words, if either the Lagrangian or the Hamiltonian does not explicitly depend on time then H is constant; it is an integral of motion. Energy conservation may be summarised as $H = T + V$.

5.3.1 Geometrical Interpretation of Hamilton's Equations

In the previous section a configuration space has been introduced, in which the n generalised coordinates are plotted and the orbit in this space corresponds to the one that is selected by the equations of motion. Now a $2n$-dimensional space is considered, in which the generalised coordinates and generalised momenta are depicted. This space is called *phase space*. The interpretation of Hamilton's equations in phase space is similar to the one for the Lagrange equations in configuration space. Through every point in phase space there is one (and only one) orbit. The progress of a phase point as a function of time is given by the variation in the Hamiltonian

$$dq_k = \frac{\partial H}{\partial p_k} dt; \quad dp_k = -\frac{\partial H}{\partial q_k} dt$$ (5.30)

Therefore, starting from a point in phase space these relations give the tangent to the orbit through that point.

5.3.2 Poisson Brackets

An arbitrary mechanical quantity $f(p_k, q_k, t)$ will change during the motion. The change is

$$\frac{df}{dt} = \sum_k \left(\frac{\partial f}{\partial q_k} \dot{q}_k + \frac{\partial f}{\partial p_k} \dot{p}_k \right) = \sum_k \left(\frac{\partial f}{\partial q_k} \frac{\partial H}{\partial p_k} - \frac{\partial f}{\partial p_k} \frac{\partial H}{\partial p_k} \right) \qquad (5.31)$$

For two arbitrary functions of the coordinates and the momenta the *Poisson bracket* of two functions f and g is defined as

$$\{f, g\} = \sum_k \left(\frac{\partial f}{\partial q_k} \frac{\partial g}{\partial p_k} - \frac{\partial f}{\partial p_k} \frac{\partial g}{\partial q_k} \right) \qquad (5.32)$$

In this way Eq. (5.31) is written as

$$\frac{df}{dt} = \{f, H\} + \frac{\partial f}{\partial t} \qquad (5.33)$$

It is the most general way of expressing the equations of motion and gives the change as a function of time of an arbitrary quantity.

The Poisson bracket has a number of useful properties. Taking for f either p_k or q_k returns the Hamilton equations. Taking H itself yields energy conservation. More generally, any quantity f that does not explicitly depend on time will be an integral of motion if $\{f, H\} = 0$. Further mathematical properties are

$$\{q_k, q_i\} = 0; \quad \{p_k, p_i\} = 0; \quad \{q_k, p_i\} = \delta_{ki} \qquad (5.34)$$

Furthermore,

$$\{f, g\} = -\{g, f\}; \quad \{f + g.h\} = \{f, h\} + \{g, h\}; \quad \{fg, h\} = \{f, h\}g + f\{g, h\} \qquad (5.35)$$

The identity

$$\{f, \{g, h\}\} + \{g, \{h, f\}\} + \{h, \{f, g\}\} = 0 \qquad (5.36)$$

is known as *Jacobi's identity*.

5.4 Canonical Transformations

It was shown that the Lagrange equations are associated with an action integral and that the choice of generalised coordinates does not really matter. The question is now whether this freedom can be employed to acquire more insight in the workings of mechanics.

Suppose there is a transformation from one set of generalised coordinates q_k, p_k to another set Q_k, P_k. The variation principles in both sets read

$$\delta \int_{t_1}^{t_2} \left[\sum_k p_k \dot{q}_k - H(p_k, q_k, t) \right] = 0; \quad \delta \int_{t_1}^{t_2} \left[\sum_k P_k \dot{Q}_k - K(P_k, Q_k, t) \right] = 0$$

$$(5.37)$$

where K is the Hamiltonian in the new generalised coordinates. The equivalence of both expressions implies that

$$\lambda \left[\sum_k p_k \dot{q}_k - H(p_k, q_k, t) \right] = \left[\sum_k P_k \dot{Q}_k - K(P_k, Q_k, t) \right] + \frac{dG}{dt} \quad (5.38)$$

Here a function G has been introduced, which is allowed as long as

$$\delta \int_{t_1}^{t_2} \frac{dG}{dt} = \delta \left[G(t_1) - G(t_2) \right] = 0 \tag{5.39}$$

G is called the *generating function*. A constant λ has also been introduced, which may be chosen to be equal to 1.

The interesting object here is the generating function. Assume for the moment that the functional dependence of G is as $G_1(q_k, Q_k, t)$. In that case

$$\sum_k p_k \dot{q}_k - H(p_k, q_k, t) = \sum_k P_k \dot{Q}_k - K(P_k, Q_k, t) + \frac{\partial G_1}{\partial t} + \sum_k \left(\frac{\partial G_1}{\partial q_k} \dot{q}_k + \frac{\partial G_1}{\partial Q_k} \dot{Q}_k \right) \tag{5.40}$$

For this to be true it is required that

$$p_k = \frac{\partial G_1}{\partial q_k}; \quad P_k = -\frac{\partial G_1}{\partial Q_k}; \quad K = H + \frac{\partial G_1}{\partial t} \tag{5.41}$$

When other assumptions about G are made, different relations for the partial derivatives follow. Four cases are distinguished:

1. type 1: $G_1(q_k, Q_k, t)$;
2. type 2: $-\sum_k Q_k P_k + G_2(q_k, P_k, t)$;
3. type 3: $\sum_k q_k p_k + G_3(Q_k, p_k, t)$;
4. type 4: $\sum_k (q_k p_k - Q_k P_k) + G_4(p_k, P_k, t)$.

All these give a relation between the two sets of canonical coordinates and momenta in terms of the generating function. In all cases it is found that

$$K = H + \frac{\partial G_{(i)}}{\partial t} \quad i = 1..4 \tag{5.42}$$

In particular the type 2 is of interest (and will be used below); when this dependence is selected it holds that

$$p_k = \frac{\partial G_2}{\partial q_k}; \quad Q_k = \frac{\partial G_2}{\partial P_k} \tag{5.43}$$

5.5   The Action Integral Revisited

The action integral has been defined as follows:

$$S = \int_{t_1}^{t_2} \pounds(\dot{q}_k, q_k, t)dt \tag{5.44}$$

Studying the variation δS, keeping the end points of the integration fixed leads to the Lagrange equations

$$\frac{d}{dt}\frac{\partial \pounds}{\partial \dot{q}_k} - \frac{\partial \pounds}{\partial q_k} = 0 \tag{5.45}$$

The geometrical interpretation in the configuration space is that there are two points, fixed at times t_1 and t_2 and the variation in the action integral sample all possible orbits between these two and the requirement of stationarity then leads to the one 'true' orbit. Another possibility is now explored. The point at t_1 is kept fixed, but a slight variation in the orbit is permitted that leads to a neighbouring point in configuration space that is an amount δq_k away. The orbits that are sampled, however, are all 'true' ones, in that they satisfy Lagrange's equations. Assuming, for illustration purposes, a one-coordinate system, varying the action integral gives

$$\delta S = \delta \int_{t_1}^{t_2} \pounds(\dot{q}, q, t)dt = \int_{t_1}^{t_2} \left(\frac{\partial \pounds}{\partial \dot{q}}\frac{d\delta q}{dt} + \frac{\partial \pounds}{\partial q}\delta q \right) dt =$$

$$= \left[\frac{\partial \pounds}{\partial \dot{q}}\delta q \right]_{t_1}^{t_2} - \int_{t_1}^{t_2} \left(\frac{d}{dt}\frac{\partial \pounds}{\partial \dot{q}} - \frac{\partial \pounds}{\partial q} \right) dt \tag{5.46}$$

The Lagrange equations hold for all orbits, so the term under the integral vanishes. Using the definition of the generalised momentum $p = \partial \pounds / \partial \dot{q}$, and setting $\delta q(t_1) = 0$ and $\delta q(t_2) = \delta q$ it is found that $\delta S = p\delta q$. If there are n generalised coordinates, more generally, it follows that

$$\delta S = \sum_{i=1}^{n} p_i \delta q_i \quad \rightarrow \quad \frac{\partial S}{\partial q_i} = p_i \tag{5.47}$$

5.5.1 The Hamilton-Jacobi Equations

From the definition of the action (5.44) it is observed that

$$\pounds = \frac{dS}{dt} \tag{5.48}$$

Combining results,

$$\pounds = \frac{dS}{dt} = \frac{\partial S}{\partial t} + \sum_i \frac{\partial S}{\partial q_i} \dot{q}_i = \frac{\partial S}{\partial t} + \sum_i p_i \dot{q}_i \;\rightarrow\; \frac{\partial S}{\partial t} = \pounds - \sum_i p_i \dot{q}_i \;\rightarrow\; \frac{\partial S}{\partial t} = -H \tag{5.49}$$

Explicitly, in terms of the coordinates and momenta

$$\frac{\partial S}{\partial t} + H\left(q_k; \frac{\partial S}{\partial q_k}; t\right) = 0 \tag{5.50}$$

This first-order partial differential equation is called the *Hamilton-Jacobi Equation*.

Returning now to the discussion on canonical transformations, it is possible to ask: what are the implications of choosing a transformation in such a way that the resulting Hamiltonian $K = 0$. This is, of course, quite an audacious choice. In that case it follows from Eq. (5.42) that

$$H + \frac{\partial G_2}{\partial t} = 0 \tag{5.51}$$

Noting the implied relation between G_2 and the transformed coordinates, especially $p_k = \partial G_2 / \partial q_k$, and it may be concluded that the generating function G_2 is just equal to the action S. The actual transformation need not even be carried out explicitly for this result to transpire.

In the coordinates in which $K = 0$ (and these are P_k and Q_k), Hamilton's equations state that $\dot{P}_k = 0$ and $\dot{Q}_k = 0$ and it follows that P_k and Q_k are constants (that is, integrals of motion). Furthermore, in the special case that the Hamiltonian does not depend explicitly on time and the system is conservative, the time-dependence of the action is given by $-Et$. Then the Hamilton-Jacobi equation takes the form

$$H\left(q_k, \frac{\partial S}{\partial q_k}\right) = E \tag{5.52}$$

5.5.1.1 Example: Mass Point in a Linear Potential Field in One Dimension

As an example, take a particle of mass m in a linear potential field—a gravity field for instance—then the Hamiltonian is

$$H = \frac{p^2}{2m} + gq \qquad (5.53)$$

The Hamilton-Jacobi equation is

$$\frac{1}{2m}\left(\frac{\partial S}{\partial q}\right)^2 + kq + \frac{\partial S}{\partial t} = 0 \qquad (5.54)$$

Setting $S = -Et + W(q)$ leads to the equation for $W(q)$

$$\frac{1}{2m}\left(\frac{\partial W}{\partial q}\right)^2 + gq = E \qquad (5.55)$$

or

$$S = \int \sqrt{2m(E - gq)}\,dq - Et \qquad (5.56)$$

Now, use that under the transformation to P and Q it holds that $Q = \partial G_2/\partial P = \partial S/\partial P$ and that Q must be constant. So, taking for P the energy E it follows that

$$Q = \int \left[\frac{m}{\sqrt{2m(E - gq)}}\right]dq - t = -\frac{\sqrt{2m(E - gq)}}{k} - t \qquad (5.57)$$

In other words

$$q = \frac{g(Q + t)^2}{2m} + \frac{E}{g}; \quad p = \frac{\partial S}{\partial q} = \sqrt{2m(E - gq)} = -g(Q + t) \qquad (5.58)$$

This result is entirely as expected, of course; however, it is a highly unusual way of getting there.

5.5.1.2 Particle in a Potential Field in Three Dimensions

This example follows trivially from the equations that have already been given. The particle has mass m and the potential field is $V(\mathbf{r}, t)$. Directly from (5.50) it is seen that

$$\frac{\partial S}{\partial t} + \frac{1}{2m}\left(\frac{\partial S}{\partial r_i}\right)^2 + V(\mathbf{r}, t) = 0 \qquad (5.59)$$

This form will be useful later on when the Hamilton-Jacobi equations are employed to make progress in quantum mechanics.

5.5.2 The Propagation of Light Waves

A surprising application of the Hamilton-Jacobi equations pertains to the propagation of light. Monochromatic light has a wave vector $\mathbf{k}$ and a circular frequency ω; the speed of light is $c = \omega/|\mathbf{k}|$. One of the components of the electrical or magnetic field is oscillatory and has the form

$$f = ae^{i(\mathbf{k}.\mathbf{r}-\omega t+\alpha)} \tag{5.60}$$

Here α is a phase factor and a is the amplitude. Strictly speaking, the real part needs to be taken, but for notation convenience that is omitted for the moment.

The *eikonal* ψ is introduced, which is just the term in the exponential

$$\psi = \mathbf{k}.\mathbf{r} - \omega t + \alpha \tag{5.61}$$

For small times and small extensions, the eikonal can be expanded

$$\psi = \psi_0 + \mathbf{r}.\frac{\partial \psi}{\partial \mathbf{r}} + t\frac{\psi}{\partial t} \tag{5.62}$$

It follows that

$$\mathbf{k} = \frac{\partial \psi}{\partial \mathbf{r}} = \mathrm{grad}\psi; \quad \omega = -\frac{\partial \psi}{\partial t} \tag{5.63}$$

Using 4-vectors $[x] = [\mathbf{r}, ict]$ and $[k] = [\mathbf{k}, i\omega/c]$, it is seen that $k^\lambda k_\lambda = 0$ (sum over $\lambda = 1...4$) and the *eikonal equation* reads

$$\frac{\partial \psi}{\partial x^\lambda}\frac{\partial \psi}{\partial x_\lambda} = 0 \tag{5.64}$$

It is remembered that the Hamilton-Jacobi equation is

$$\mathbf{p} = \frac{\partial S}{\partial \mathbf{r}}; \quad H = -\frac{\partial S}{\partial t} \tag{5.65}$$

Compare this with (5.63) and the correspondence $\mathbf{k} \to \mathbf{p}, \omega \to H, S \to \psi$ becomes evident.

Hamilton's equations are

$$\dot{\mathbf{p}} = -\frac{\partial H}{\partial \mathbf{r}}; \quad \mathbf{v} = \dot{\mathbf{r}} = \frac{\partial H}{\partial \mathbf{p}} \tag{5.66}$$

Applying the correspondence, it follows that

$$\dot{k} = -\frac{\partial \omega}{\partial r}; \quad \dot{\mathbf{r}} = \frac{\partial \omega}{\partial \mathbf{k}} \tag{5.67}$$

In vacuum $\omega = c|k|$ (independent of the position), so

$$\dot{k} = 0; \quad \mathbf{v} = \dot{\mathbf{r}} = c\mathbf{n} \tag{5.68}$$

where $\mathbf{n}$ is the unit vector in the direction of the wave vector.

It may therefore be concluded that light moves in a straight line with speed c.

The correspondence identified here will be useful when the Hamilton-Jacobi equations are used in quantum mechanics. Newton thought of light as a stream of particles, while later on physicists, describing, for example, interference phenomena, concluded that it must be wave-like in character. The fact that there is a correspondence, as the Hamilton-Jacobi formalism shows, implies that the two views can be reconciled. The analysis above started out as a wave-like phenomenon and by showing the correspondence particle mechanics appears to be applicable. Modern physics starts here.

References

Kibble TWB (1985) Classical mechanics, 3rd edn. Longman Scientific and Technical, Harlow
Landau LD, Lifshitz EM (1976) Mechanics (Course on theoretical physics, Volune 1). Pergamon, Oxford

Chapter 6
Special Relativity Theory

Abstract The concept of an inertial frame and the implications of the Michelson-Morley experiment are discussed. Because of its great importance to physics in the twentieth century an historical narrative against the background of Maxwell's equations is presented. Then the Lorentz transformation is derived. Its consequences for space-time phenomena—Lorentz contraction and time dilation—are demonstrated. The momentum-energy four-vector is introduced and relativistic collisions are treated. The velocity and acceleration of four-vectors are briefly mentioned and the preservation of the link with Newtonian mechanics is emphasised. Minkowski space and world lines are graphical aids and a more general discussion of four-vectors and their transformation rules are expounded. An accelerated system is analysed, which shows the concept of an 'event horizon'. Electromagnetic theory in the context of relativity theory and its relation to Maxwell's equations the continuity equation is treated. A number of examples are given, one of which is the resulting solution for light waves.

6.1 Inertial Frames

A discussion of relativity theory should start with the most basic assumption of Newtonian mechanics, which is Newton's first law (strictly speaking due to Gallileo): 'A body persists in its motion unless a force is exerted on it'. A coordinate frame may be fixed to such a body (that is one that feels no forces) and—relative to the fixed stars—that coordinate frame will move with the body and maintain a constant velocity. Another body, which also moves at a constant velocity with respect to the fixed stars, may have a relative velocity compared to the first body and, again, a coordinate frame may be attached to it. In fact, there are an infinite number of frames, all distinguished by their relative velocity, in which there are no forces and hence no accelerations. Such frames are called *inertial frames*. Special relativity deals with the interrelation of physical phenomena between inertial frames.

Suppose that an electric charge is carried in one frame. In that frame, the charge stands still and the physicist who performs measurements in that frame will detect an electric field. The effects of that same charge measured by a physicist in another frame, one that moves with a relative velocity compared to the one in which the charge

M. A. C. Koenders, *Constructing the Edifice of Mechanics*, Undergraduate Texts in Physics, https://doi.org/10.1007/978-3-031-34071-0_6

stands still, will not only be an electric field but also a magnetic field. Similarly, a measurement of momentum in the frame in which a body stands still will be zero, but the momentum as measured in a neighbouring frame—one that zooms past with a relative velocity V—will be—mV. So, it is concluded that the outcome of an experiment depends on the frame in which it is measured; there is nothing mysterious about that. For the physicists doing the measurements in the frames, however, there is no way of telling which frame they are in—they do not feel a force, there is no acceleration, and they are in inertial frames; however, by exchanging signals, they can figure out what their *relative velocity* is.

In Chap. 1, it was seen that the relationship between time and spatial quantities in two inertial frames that move with constant relative velocity V is given by the Galileo transformation. In Fig. 6.1, the idea is illustrated in the so-called *standard configuration*. Observers located in the origins of both frames carry clocks that are synchronised such that at time $t' = t = 0$ the origins coincide. For simplicity, the x_1 coordinates are chosen to align, so the other coordinates remain invariant.

Now, consider an event that takes place at location $\mathbf{x}$ for the unprimed observer and at location $\mathbf{x}'$ for the primed observer. The answer is simple: $x_1' = x_1 - Vt$, $x_2' = x_2$, $x_3' = x_3$ and $t' = t$. Also, $v_1' = v_1 - V$. As noted in Chap. 1, this is the classical transformation and embedded in it is the idea of *universal time*, that is all clocks—no matter what the inertial frame they are in—(and assuming that they are all manufactured to the same standard of precision) run synchronously. This was essentially the view of physics at the end of the nineteenth century. Then problems became apparent. It is not essential to read the history of physics to understand relativity theory, but it helps to gain an insight into the enormous intellectual effort that was required to take mankind from a Newtonian world view to the concepts that rule modern physics. It must be emphasised that Newtonian physics is still very valid, but its validity is confined to a certain parameter range, which shall be discussed in some detail below. As noted before in Chap. 1, a *classical limit* can be taken to transform the theory that holds more generally back to the comfort of the Newtonian habitat.

Fig. 6.1 Inertial frames

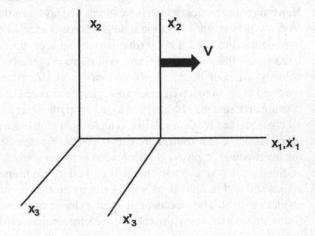

6.2   A Brief Historical Note

In 1873, Maxwell published his *Treatise on Electricity and Magnetism*, Maxwell (1873), in which he put forward the unified equations that are now known as the Maxwell Equations. These will be looked at in some detail below in Sect. 6.7.3. For the moment though, looking at these equations in a vacuum, there are a couple of very striking observations that can be made. First, the time appears explicitly and second there is a velocity in these equations that appears as a constant (the latter turns out to be the speed of light c). The big question at the time was then: is this velocity the same everywhere, or does an observer who is, for example, at the Equator measure the same velocity—and thereby the same behaviour of electromagnetic fields—as one at the North Pole? The latter will have a certain velocity with respect to the former. So, what happens?

Maxwell (1831–1879) is the first important name in the story. The other names that are central to the development are Michelson (1852–1931), Morley (1838–1923), Lorentz (1853–1928) and Einstein (1879–1955).

Maxwell's equations yield the electromagnetic fields when the configuration of current density and charge density are given. In the absence of charges and currents, these equations also have solutions, which may be interpreted as light waves. The velocity of these is a function of the electromagnetic properties of the vacuum, which can be measured in static systems. The numerical agreement with the velocity of light was a great triumph for Maxwell's theory. (Maxwell's equations are anything but simple, and it took a while before a little more insight into the structure of the solutions was acquired—especially through the work of Hertz (1857–1894)). The electromagnetic waves were believed to be transmitted through an 'aether', much like sound waves in air. In an experiment that is at rest with respect to the aether a certain value for the speed of light is found, while physicists who travel with a velocity with respect to the aether would find a different one, one that would also depend on the orientation. In principle then, one would conclude that the constant c in Maxwell's equations was different from one inertial frame to the next; in addition, it would be direction-dependent. This view makes the speed of light also direction-dependent, something that could be tested experimentally. In 1881, Michelson carried out an experiment to verify this. The speed of light was measured in two different directions, normal to each other. No difference was found. However, Michelson made a mistake in his calculations and, on re-calculation, it was determined that the error in his measurement was of the same order of magnitude as the order of magnitude of the velocity of the Earth with respect to the Sun (while one might roughly assume that the aether was at rest with respect to the solar system, no living soul believed that the aether was at rest with respect to the Earth, or—in other words—that it would turn around the Sun!). The results of a second, much more accurate set of experiments was published in 1887 by Michelson and Morley. No effect was observed! The speed of light was direction-independent.

The experimentalists having had their say, now it was the turn of the theoreticians to grapple with the subject matter. A key figure in the debate around the speed of light and the notion of an aether was Lorentz. (Lorentz was based in Leyden in the Netherlands, spoke four languages fluently and was one of the greatest scientific communicators of his time). In 1892, he put forward the idea that the length of a material object may be dependent on the motion of that object with respect to the aether. In 1904, (in part responding to the idea, put forward by Poincaré, that there is such a thing as the 'principle of relativity') he published his theory *Electromagnetic Phenomena in a System Moving with Any Velocity Less Than Light*, Lorentz (1904). He is now very close to some version of relativity theory and the work contains the Lorentz transformation, which—as will be seen below—is central to the subject matter. He grapples with the problem of time measurement and, indeed, comes up with time dilation by introducing a 'local time', a quantity that cannot be measured with a traditional clock, but rather a parameter in the theory. He cannot abandon the concept of universal time as was supposed in Newtonian mechanics. Lorentz' insights were very much based on his concept of the vacuum, which he sees as a space entirely filled with charged particles. He *does*, however, arrive at the apparent mass dependence on the motion of a body. It is possible that given time, he would have realised that the concept of universal time is wrong, but the intellectual stamina that a step of such magnitude required did not materialise.

A number of other theoreticians were involved. For example, Voigt (1850–1919), who in 1887 derived the Voigt transformation, which is the transformation that leaves Maxwell's equations in vacuum (where there are no charges and no currents) invariant; this turns out to be the Lorentz transformation.

Then there was Reynolds (1842–1912) who wrote a number of papers on densely packed granular materials (these exhibit the strange property of dilatancy, see Koenders (2020)) and put forward the concept of the vacuum as consisting of such a material, basically a kind of sand. He could then derive the speed of light, by assuming elastic properties of the grains. In 1902, he published a massive book entitled *On the Sub-Mechanics of the Universe*, Reynolds (1903), in which he explained his ideas, which are entirely classical and add nothing to the subject matter. However, Reynolds was at the time highly regarded and a famous scientist who did groundbreaking work in fluid mechanics and granular mechanics and it shows how difficult it was—in an intellectual sense—to let go of the Newtonian world view and also of the great need to explain the properties of light transmission in a vacuum and say goodbye to the aether.

It is then all the more impressive to consider the work of Albert Einstein, who—at the age of 25—in 1905 published the paper, Einstein (1905a), that lays the foundation for relativity theory and in which he radically abandoned the notions of absolute time, absolute space and the aether. A year later the second paper came out, detailing the relativistic interpretation of momentum and energy, the famous $E = mc^2$ paper, Einstein (1905b). These two papers, both very simple with only a handful of equations, shook the world of physics to its core. Newtonian mechanics became merely a classical limit, useful only in certain parameter ranges. A startling note is that Einstein was not aware of Lorentz's 1904 publication.

The development of relativity theory did not stop in 1906. One of the issues that became very urgent was the reformulation of Newton's universal gravity, which Einstein also did (together with the mathematician Grossmann) in various stages in 1913–1918: Einstein (1913, 1918), Grossmann (1913). The consequences of that theory are numerous: black holes, a more or less plausible model of the expanding universe and gravitational radiation are all aspects that the physics and astronomy community (both theoreticians and experimentalists) are still grappling with, see Weinberg (1972).

The late nineteenth and early twentieth century was a period when mankind suddenly made enormous intellectual progress. In physics, other than relativity theory, the first steps on the road to quantum mechanics were taken. Freud wrote his famous paper on the interpretation of dreams and Schönberg began experimenting with radically new musical forms. Perhaps it was only in the 'zeitgeist' of that period that a young Einstein could blow away the cobwebs of Newtonian physics and propose such radical ideas as relative time and space. The theory that deals with gravity in a relativity context is known as *General Relativity Theory*; it requires space-time to be curved in the vicinity of mass (and does away with Newton's gravitation law other than a classical limit). Its mathematical framework is Riemannian geometry, which enables partial differentiation in a non-Euclidean space. The much simpler theory, which excludes gravitational effects and leaves space Euclidean is called *Special Relativity Theory*. The latter theory will be treated below and it also, quite naturally, includes electro-magnetic theory conceived by Maxwell, which—as may be seen from the brief historical overview above—obviously plays a key role in the foundations of this subject.

6.3 The Lorentz Transformation

The Michelson-Morley experiment is interpreted as follows: independent of the inertial frame the speed of light in vacuum, c, always has the same value. This is the starting point of Relativity Theory. Now, consider two inertial frames that move with velocity V with respect to one another. Observers are positioned in either frame. They carry clocks and measuring rods that have been manufactured to the exact same specification and precision. The clocks are synchronised at the point when the two origins coincide. A beam of light is then emitted from the origin to a point in space P (see Fig. 6.2). Measured from the unprimed frame the time t the light takes to travel to P is given by

$$x_1^2 + x_2^2 + x_3^2 - c^2t^2 = 0 \tag{6.1}$$

Similarly, the very same as viewed from the primed frame a time t' is given by

$$x_1'^2 + x_2'^2 + x_3'^2 - c^2t'^2 = 0 \tag{6.2}$$

Using the two equations above, it may be concluded that

$$x_1^2 + x_2^2 + x_3^2 - c^2t^2 = K\left(x_1'^2 + x_2'^2 + x_3'^2 - c^2t'^2\right) \tag{6.3}$$

where K is a constant.

Fig. 6.2 Inertial frames with a light beam going from the origin to P

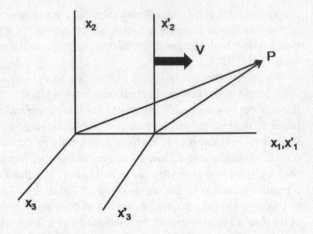

The constant is rather arbitrary, as there is no preference of the primed framed to the unprimed frame and so, if the constant has any physical meaning, it must be something that prejudices either frame. No such prejudice is evident. It must therefore be concluded that $K = 1$.

A more transparent discussion is provided by introducing a fourth coordinate, in addition to the three spatial coordinates. This may be done by setting $x_4 = ict$ and $x_4' = ict'$. Components of these 'four vectors' are denoted by a Greek subscript; the result of the experiment with a light beam going from the origin to P is then summarised as

$$x_\mu x_\mu = x_\nu' x_\nu' \tag{6.4}$$

(Einstein's summation convention has been applied).

Essentially what this expresses, is that the length of the four-vector is invariant. Proposing then a linear transformation—and excluding the unphysical case of reflection—a rotation is put forward. The transformation from the unprimed to the primed frame takes the form

$$x_\mu' = L_{\mu\nu} x_\nu \tag{6.5}$$

The transformation L must have the property that $\det(L) = 1$, as it is a rotation.

A special case, as represented in Fig. 6.2, is the one in which the x_2 and x_3 components are left invariant and the components x_1 and x_4 transform as a plane rotation over an angle ϕ. Because the fourth component is imaginary, the angle must be imaginary

$$\begin{bmatrix} x_1' \\ x_2' \\ x_3' \\ x_4' \end{bmatrix} = \begin{bmatrix} \cos\phi & 0 & 0 & -\sin\phi \\ 0 & 1 & 0 & 0 \\ 0 & 0 & 1 & 0 \\ \sin\phi & 0 & 0 & \cos\phi \end{bmatrix} \begin{bmatrix} x_1 \\ x_2 \\ x_3 \\ x_4 \end{bmatrix} \tag{6.6}$$

Introduce the auxiliary variable $v = ic \tan \phi$; then it follows that

$$\cos \phi = \frac{1}{\sqrt{1 - v^2/c^2}} \text{ and } \sin \phi = -\frac{iv/c}{\sqrt{1 - v^2/c^2}} \qquad (6.7)$$

Using that $x_4 = ict$, $x_4' = ict'$, it is seen that

$$\begin{aligned}
x_1' &= \frac{1}{\sqrt{1 - v^2/c^2}} (x_1 - vt) \\
x_2' &= x_2 \\
x_3' &= x_3 \\
t' &= \frac{1}{\sqrt{1 - v^2/c^2}} \left(t - \frac{v}{c^2} x_1 \right)
\end{aligned} \qquad (6.8)$$

A first-order Taylor expansion in v/c, that is $v \ll c$, yields the classical limit

$$x_1' = x_1 - vt; \ t' = t \qquad (6.9)$$

This is just the Galileo transformation with $V = v$. So, the auxiliary variable v may be interpreted as the relative velocity with which the frames move alongside one another.

The transformation L is called the Lorentz transformation and has the structure of a rotation in four dimensions. In the Galileo transformation, because of the condition $t' = t$, space and time can be treated as independent, but in relativity theory time and space cannot be treated as separate entities; transformations need both space and time coordinates—they are four-vectors. Note that the effects of relativity theory become noticeable only when velocities approach the speed of light. It can also be seen (due to the factors $\sqrt{1 - v^2/c^2}$) that no speeds higher than the speed of light are physically possible. Both these concepts, the intertwining of space and time and the existence of a maximum speed in nature, have caused humanity much anguish and to this day many find it intuitively unacceptable (only in science fiction is superluminal communications treated seriously). However, the constancy of the speed of light in all inertial frames, as confirmed by experiment, leads to these conclusions and unless Nature says otherwise, it has to be accepted.

Einstein's paper in 1905, Einstein (1905a), a year after Lorentz's book (Lorentz 1904) appeared (which he had not yet read), and in two and a half pages contains only seven equations. By all accounts, this is a simple paper, but it had a massive influence on the world, the further development of physics and also on philosophy; it turned Einstein into a scientific superstar. Being Jewish, he could not remain in Berlin in the 1930s. He emigrated to the US, where he was employed at Princeton University.

6.4 Lorentz Contraction and Time Dilation

The consequences of the Lorentz transformation are many, but two very obvious ones are discussed here. There are many variations on these two examples, but they all come down to the same thing: time and space need to be treated not independently—

as in Newtonian physics—but *together* in four-vectors. In addition, the rather specific way in which measurement is prescribed in relativity theory must be noted.

6.4.1 Lorentz Contraction

This aspect really deals with the measurement of a length. As long as a length is measured in the frame in which the measuring rod stands still, there is no problem. The question is: what does an observer who moves along in another frame find? To that end two signals are emitted (in the unprimed frame): one from the origin at time $t = 0$, which corresponds to the four-vector

$$\begin{pmatrix} 0 \\ 0 \\ 0 \\ 0 \\ 0 \end{pmatrix} \tag{6.10}$$

Another signal is emitted at time $t = \tau$ and at a distance $x_1 = \lambda$. This event corresponds to the four-vector

$$\begin{pmatrix} \lambda \\ 0 \\ 0 \\ ic\tau \end{pmatrix} \tag{6.11}$$

Using the Lorentz transformation, these two signals are received in the primed frame as

$$\begin{pmatrix} 0 \\ 0 \\ 0 \\ 0 \\ 0 \end{pmatrix} \quad \text{and} \quad \frac{1}{\sqrt{1 - v^2/c^2}} \begin{pmatrix} \lambda - v\tau \\ 0 \\ 0 \\ -iv\lambda/c + ic\tau \end{pmatrix} \tag{6.12}$$

The 'length measurement' in the primed frame is carried out in such a way that the two signals arrive at the same time, in other words, choose $\lambda = c^2\tau/v$. The length, so measured, in the primed frame is then

$$\frac{1}{\sqrt{1 - v^2/c^2}} (\lambda - v\tau) = \lambda \sqrt{1 - v^2/c^2} \tag{6.13}$$

Shorter, therefore, than the length in the unprimed frame. However, the outcome depends crucially on the choice for the value of λ; as noted above, the measurement methodology has to be specified very carefully. If one is sloppy with this, all sorts of weird conclusions can be arrived at.

6.4.2 Time Dilation

In some sense, time dilation is easier to calculate than Lorentz contraction. Two signals are emitted from the unprimed frame, one at time $t = 0$ from the origin and the other at time $t = \tau$, also from the origin. The two four-vectors are therefore

$$\begin{pmatrix} 0 \\ 0 \\ 0 \\ 0 \\ 0 \end{pmatrix} \text{ and } \begin{pmatrix} 0 \\ 0 \\ 0 \\ 0 \\ ic\tau \end{pmatrix} \tag{6.14}$$

Transforming these two four-vectors results in

$$\begin{pmatrix} 0 \\ 0 \\ 0 \\ 0 \\ 0 \end{pmatrix} \text{ and } \frac{1}{\sqrt{1 - v^2/c^2}} \begin{pmatrix} -v\tau \\ 0 \\ 0 \\ i\tau \end{pmatrix} \tag{6.15}$$

So, it is concluded that the signals do not arrive at the same time interval in the primed frame. The time interval has been lengthened by a factor $1/\sqrt{1 - v^2/c^2}$. Of course, the two signals do not arrive at the same position in the primed frame.

The effect of time dilation has been observed in fast-travelling elementary particles, for example, the μ-meson. At rest, these particles have a mean lifetime of 2.199×10^{-6} s. For moving particles, longer values are found, depending on their velocity, exactly as relativity theory predicts.

The *twin paradox*, or *clock paradox* is based on time dilation. The idea is that one of the twins goes off, travelling fast into deep space, with his clock (that could be a biological clock), while the other one stays at home, also with his clock. After some long journey, the space traveller turns around very quickly, which involves a brief spell of very high acceleration, and travels back until he meets his twin again. Note that time dilation does not depend on the sign of the velocity, as it is quadratic in v. Two points of view may be taken. The first is from the one who stays at home; he notices his brother travelling off at velocity v and coming back at velocity $-v$ and therefore concludes that the traveller's clock is slow by the dilation factor. The travelling twin notices the same thing about his stay-at-home sibling, because he sees him go off at $-v$ and come back at v. So, when they are reunited, who is older? That is the paradox. The reasoning is flawed however, because the travelling twin needs to turn around and travel back and must therefore experience accelerations and—even for a short while—is then not in an inertial frame.

6.5 Momentum and Energy

As noted in the introduction to this chapter, relativity theory should say something about momentum. A measurement of momentum in the frame in which a body stands still will be zero, but the momentum of that body as measured in a neighbouring frame—one that zooms past with a relative velocity V—will be $-mV$. The Lorentz transformation works on four-vectors, while traditional momentum is, of course, a three-component vector. In addition, momentum as understood in classical mechanics, which for a point mass is $m\mathbf{v}$, need not have exactly the same form in the context of relativity theory. However, in the classical limit $v \ll c$ it should converge to it.

Assume then that a particle at rest has zero momentum. In that case the four-vector that represents it must have three zero components. The fourth component is a bit of a mystery, but it must exist as on transforming to a moving frame the first three components must acquire a value. The idea is now that a particle at rest is represented by a four-vector of the form

$$\begin{bmatrix} 0 \\ 0 \\ 0 \\ 0 \\ iX \end{bmatrix} \tag{6.16}$$

Applying a Lorentz transformation to this four-vector will give something that, when the classical limit is forced, should look like $-mv$ (for simplicity a motion in the x-direction is assumed; this can always be made more general by performing an 'ordinary' spatial rotation). The transformed vector is

$$\frac{1}{\sqrt{1 - v^2/c^2}} \begin{bmatrix} -vX/c \\ 0 \\ 0 \\ 0 \\ iX \end{bmatrix} \tag{6.17}$$

In the classical limit, up to second order in v/c, this becomes

$$\begin{bmatrix} -Xv/c \\ 0 \\ 0 \\ 0 \\ iX \left(1 + \frac{1}{2}v^2/c^2\right) \end{bmatrix} \tag{6.18}$$

This would suggest that the choice $X = mc$, leads to the correct value of the classical momentum.

The fourth component can be further elucidated. In the classical limit, it takes the form $i \left(mc + \frac{1}{2} mv^2/c \right)$. The classical kinetic energy is $\frac{1}{2} mv^2$ and therefore, but for a factor of ic, the fourth component represents an energy. The relativistic interpretation is now that at rest mass points possess a 'rest energy' mc^2. In motion with velocity v the energy becomes $(1 - v^2/c^2)^{-1/2} mc^2$. In the classical limit, $v \ll c$, this becomes $mc^2 + \frac{1}{2} mv^2$.

The above reasoning, while persuasive, is entirely speculative. Basically, it is *assumed* that momentum transforms from one frame to another via a Lorentz transformation. The transformation, which has been derived for a space-time four-vector, has been extended in validity to apply to momentum-energy. Such a procedure is not necessarily allowed, but must be underpinned by experiments. One experiment that could be done and which goes to the heart of the concept of momentum is a *collision experiment*. At the same time, the energy conservation in an elastic collision can be evaluated. In addition to the unprimed and primed coordinate systems, a double-primed system is introduced. Let this double-primed system move with velocity v with respect to the primed system and let the latter move with velocity w with respect to the unprimed system. The first question to ask is: if the transformation from the double-primed system to the primed system is the Lorentz transformation $L(v)$ and the transformation from the primed to the unprimed system $L(w)$, what is the transformation from the double-primed system to the unprimed system. To that end, calculate the product $L'' = L(w)L(v)$. It is easily worked out that the result is again a Lorentz transformation $L'' = L(u)$, with u given by the formula

$$u = \frac{v + w}{1 + vw/c^2} \tag{6.19}$$

(It is obviously not $v + w$, as this would enable velocities greater than the speed of light.) This is easy to understand when it is recalled that the Lorentz transformation represents a rotation over an angle ϕ and the addition of two tangents is simply

$$\tan(\phi_1 + \phi_2) = \frac{\tan \phi_1 + \tan \phi_2}{1 - \tan \phi_1 \tan \phi_2} \tag{6.20}$$

A sample collision experiment is described in the unprimed system as a particle with mass m and velocity v (only the x-component is used), colliding with another particle with mass $2m$ and velocity $-v/2$. In classical terms, the momentum before the collision equals 0. After the collision, which is assumed to be entirely elastic, the first particle has velocity $-v$ and the second particle $v/2$. Momentum is thus conserved. Now, let another system, the primed system move with a velocity w with respect to the unprimed system. Using the formula for the addition of velocities (6.19), the before and after momenta appear to be

$$m\frac{v + w}{1 + vw/c^2} + 2m\frac{-v/2 + w}{1 - vw/(2c^2)} \text{ and } m\frac{-v + w}{1 - vw/c^2} + 2m\frac{v/2 + w}{1 + vw/(2c^2)} \tag{6.21}$$

Using a few simple numbers, it is easily shown that momentum, so defined, is *not* conserved.

It will now be shown that momentum, as defined by Eq. (6.18) (with $X = mc$ and $2mc$, respectively) *is* conserved. To that end start with the four-vectors in their rest systems, these are

$$p^{(a)} = \begin{bmatrix} 0 \\ 0 \\ 0 \\ imc \end{bmatrix} \text{ and } p^{(b)} = \begin{bmatrix} 0 \\ 0 \\ 0 \\ 2imc \end{bmatrix} \tag{6.22}$$

In the unprimed system, the momenta before and after the collision take the form

$$\text{before: } p^{(1)} = L(v)p^{(a)} \text{ and } p^{(2)} = L(v')p^{(b)}$$
$$\text{after: } q^{(1)} = L(-v)p^{(a)} \text{ and } q^{(2)} = L(-v')p^{(b)} \tag{6.23}$$

A particular choice for v' has been made, such that for the first three components $p^{(1)} + p^{(2)} = 0$, which does not mean to say that $v' = -v/2$.

Conservation of momentum implies

$$q^{(1)} + q^{(2)} = p^{(1)} + p^{(2)} \tag{6.24}$$

This bit of shorthand represents four equations. In the primed system, the momenta become

$$\text{before: } p'^{(1)} = L(w)p^{(1)} \text{ and } p'^{(2)} = L(w)p^{(2)}$$
$$\text{after: } q'^{(1)} = L(w)q^{(1)} \text{ and } q'^{(2)} = L(w)q^{(2)} \tag{6.25}$$

Now,

$$p'^{(1)} + p'^{(2)} - q'^{(1)} - q'^{(2)} = L(w)\left(p^{(1)} + p^{(2)} - q^{(1)} - q^{(2)}\right) = 0 \tag{6.26}$$

So, if momentum is conserved in the unprimed system, it is also conserved in the primed system. Note that the fourth component is also conserved, that is to say, the quantity $m/\sqrt{1 - v^2/c^2}$ is also a conserved quantity. In classical mechanics, the kinetic energy is a conserved quantity in collisions. In relativity theory, the quantity $mv^2/2$ is not conserved, but $m/\sqrt{1 - v^2/c^2}$ is. All these things have been tested in experiments and the procedure outlined here for constructing the momentum-energy four-vector is doubtlessly correct.

The implications of the theory are far-reaching. By way of example, take the special case of light. Light is known to have energy and momentum (in an experiment letting light fall on mirror results in a rebound on the mirror, as evidenced by the 'light mills' (officially known as a *Crookes radiometer*), in which a device with an almost friction-free spindle and vanes that are shiny on one side and light-absorbing black on the other, will start spinning when illuminated).

The light four-vector may be described by taking the limit in the fourth component of $v \to c$, while $m \to 0$ in such a way that p_4 remains constant: $p_4 = iE/c$. Then the three-momentum is $\mathbf{p} = E\hat{\mathbf{v}}/c$, with $\hat{\mathbf{v}}$ the unit vector in the direction of travel of the light beam. A mental image of this limiting case is a particle with no mass, moving with the speed of light.

Consider an atom at rest. Assume that the atom emits two beams of light, each with energy E, going in opposite directions. The three-momentum of the light is then zero. The fourth component of the total momentum is conserved, so if before the emission the atom has a mass m and after the emission a mass m', the conservation law requires

$$mc^2 = E + E + m'c^2 \text{ or } m' = m - 2E/c^2 \qquad (6.27)$$

The mass of the atom has thus decreased and mass has been converted into energy. The maximum amount of mass that can be converted is the mass of the atom itself, so this mass represents a store of energy of magnitude

$$E = mc^2 \qquad (6.28)$$

It is not necessary to use light as an energy carrier. Any process whereby particles decay into other particles may exhibit an energy surplus. Nuclear fission of a suitable uranium or plutonium isotope is an example.

6.6 Velocity and Acceleration

The momentum-energy four-vector of a particle that moves with velocity $\mathbf{v}$ is $p = (1 - v^2/c^2)^{-1/2}[m\mathbf{v}, imc]$ is proportional to the mass of a particle m. Dividing by m gives a new four-vector, which plays the role of a velocity and will be called u, with components u_μ, $\mu = 1...4$. This is the relativistic generalisation of the velocity, see also Rindler (1966). The idea is that the velocity can be obtained in two ways. The first is by a Lorentz transformation from a frame in which the particle is at rest; the four-vector at rest is $[\mathbf{0}, ic]$. The second way is by differentiation of a distance with respect to time, which served so well in classical mechanics. To reconcile the second approach with the first, it has to be stated precisely what is meant by 'time'.

Let the particle move an infinitesimal amount $d\mathbf{s}$ in a time dt. The four-vector inner product is $ds^2 = d\mathbf{s}.d\mathbf{s} - dt^2$. The inner product is invariant under a Lorentz transformation and in particular in the system in which the particle is at rest it holds that $ds^2 = -d\tau^2$, where τ is the time in the frame that is co-moving with the particle, the so-called 'rest frame'; it is the particle's own time—its *proper time*. The differentiation with respect to time alluded to above is unambiguously defined by differentiating with respect to the proper time

$$u_\mu = \frac{dx_\mu}{d\tau} \qquad (6.29)$$

A small time interval in a system that moves with respect to the rest frame with velocity $\mathbf{v}$ is $dt = d\tau/\sqrt{1 - v^2/c^2}$. In passing it is noted that $u_\mu u_\mu = -c^2$.

The concept of acceleration can be treated in a similar manner and is required—and useful—to preserve Newton's second law in some form or other. There is no need to deviate very much from the classical form of Newton's law, but care must be taken to specify exactly in which reference frame an operation is carried out.

It was seen that the four-velocity is obtained from the four-position by differentiating with respect to the time as measured in the rest frame τ. A four-acceleration can in this way be obtained by doing the same again so that in the rest frame $a_\mu = dv_\mu/d\tau = d^2x_\mu/d\tau^2$. Using the identity $u_\mu u_\mu = -c^2$, it follows by differentiation that $u_\mu a_\mu = 0$ so that the four-acceleration is normal to the four-velocity. The four-momentum was $p_\mu = mu_\mu$. The first question is then: what form does the law of inertia take in relativity theory? Postulate the following law of inertia

$$\frac{dp_\mu}{d\tau} = 0 \qquad\qquad (6.30)$$

It follows that $ma_\mu + u_\mu dm/d\tau = 0$; multiply with u_μ and sum over μ yields $-c^2 dm/d\tau = 0$; therefore m is a constant. Consequently, $ma_\mu = 0$, so the inertial law is Lorentz-invariant—in other words, it holds in every inertial system. It is easy to see that the classical limit of the relativistic inertial law is obtained by considering the spatial part (remember $dt = \gamma d\tau$, with $\gamma = 1/\sqrt{1 - v^2/c^2}$): from $d\mathbf{p}/d\tau = 0$, it follows that $d\mathbf{p}/dt = 0$, which is the classical inertial law.

The spatial part of the equation of motion in the particle rest system is $d\mathbf{p}/d\tau = \mathbf{F}$. Generalising this to the relativistic case is the so-called *Minkowski equation*

$$\frac{dp_\mu}{d\tau} = F_\mu \text{ or } ma_\mu = F_\mu \qquad\qquad (6.31)$$

where F is the four-force. The spatial part of F is $\gamma\mathbf{F}$ and the spatial part of $dp/d\tau$ is $d\mathbf{p}/d\tau = \gamma d\mathbf{p}/dt$ so that the first three components of Minkovsky's equation read $d\mathbf{p}/dt = \mathbf{F}$, which is exactly as expected. What about the fourth component F_4? Use $u_\mu F_\mu = mu_\mu a_\mu = 0$ and therefore $[\gamma\mathbf{u}, i\gamma c][\mathbf{F}, F_4] = 0$, or $F_4 = i\gamma\mathbf{F}.\mathbf{u}/c$. The fourth component of the Minkowski equation is then $d(m\gamma c^2)d\tau = \gamma\mathbf{F}.\mathbf{u}$, or equivalently, $d(\gamma mc^2)/dt = \mathbf{F}.\mathbf{u}$, which is just the rate of working on the particle.

6.6.1 World Lines and Four-Vectors and Tensors

6.6.1.1 World Lines

The path of a particle can be plotted in the (x, ct) plane. The y and z directions are normal to x. The plot is called a *world line*. The space (x, y, z, ct) is called the *Minkowski space*, see also Rindler (1966). Light travels either forward in space or backwards in space, but always forward in time. In Fig. 6.3 the solid lines represent

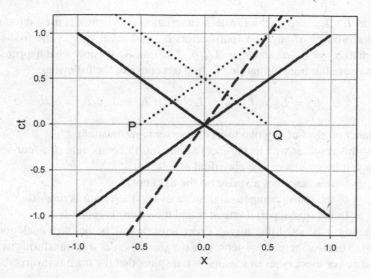

Fig. 6.3 World lines of light (solid lines) either travelling in the positive or negative direction; world line of a particle at a velocity of $c/2$ (dashed line) and world lines of light originating either in point P or point Q (dotted lines)

the paths of light travelling in the plus or minus x-direction. Lines that have angle greater than 1 or less than -1 would go faster than the speed of light and are not physically possible. A particle travelling at a velocity $v < c$ has a representation in this diagram that looks like the dashed line in Fig. 6.3. It is also a straight line.

If the y direction is included, the bundle of lines that represent light form a cone. This is called the *light cone*. Points that have $ct > 0$ point to the future; those that have $ct < 0$ come from the past. Points outside the light cone at $ct = 0$ could be said to be 'elsewhere'.

The question is now whether a particle travelling through space can 'see' the whole space, or if there are regions that cannot be surveyed. By taking any point—for example, P in Fig. 6.3—and constructing a light world line through it (the dotted lines in Fig. 6.3), it is seen that at some stage the world line of the light and the world line of the particle cross and so the light reaches the particle. The light does not have to travel in the positive x-direction; it could be beamed backwards. The light coming from point Q is such a case.

The points do not necessarily originate from $t = 0$ either. Any 'event' characterised by (x, ct) can be observed by the observer who travels at constant velocity in an inertial frame. This 'observability' is an associated property of motion in inertial frames.

6.6.1.2 Four-Vectors

It was seen in the previous sections that the Lorentz transformation is fundamentally a rotation. That implies that its inverse is its transposed: $L_{\mu\nu}^{-1} = L_{\nu\mu}$ (the inverse is defined as $L_{\mu\nu}^{-1}L_{\nu\epsilon} = \delta_{\mu\epsilon}$; the Einstein summation convention is used throughout). The inner product of any two four-vectors a_μ and b_ν, is $a_\mu b_\mu$. The four-vectors transform as $a'_\mu = L_{\mu\nu}a_\nu$ and $b'_\mu = L_{\mu\epsilon}b_\epsilon$. The question is now what happens to the inner product after transformation. To find out consider the following:

$$a'_{mu}b'_\mu = L_{\mu\nu}a_\nu L_{\mu\epsilon}b_\epsilon = L_{\epsilon\mu}^{-1}L_{\mu\nu}a_\nu b_\epsilon = \delta_{\epsilon\nu}a_\nu b_\epsilon = a_\nu b_\nu \qquad (6.32)$$

So the inner product of any two four-vectors remains invariant.

Using this result as well as the fact that the fourth component of a four-vector is imaginary, four-vectors can be classified as follows:

$a_\mu a_\mu = 0$: this describes a vector on the light cone.

$a_\mu a_\mu < 0$: the fourth component dominates and the vector is *time-like*.

$a_\mu a_\mu > 0$: the spatial part dominates and the vector is *space-like*.

In passing it is noted that higher order tensors may be defined, analogously to 3-tensors. So, for example, a 4-tensor of the second-order $a_{\mu\nu}$ transforms as $a'_{\mu\nu} = L_{\mu\epsilon}L_{\nu\zeta}a_{\epsilon\zeta}$. For these types of tensors, it transpires that the trace is invariant:

$$a'_{\mu\mu} = L_{\mu\epsilon}L_{\mu\nu}a_{\epsilon\nu} = L_{\epsilon\mu}^{-1}L_{\mu\nu}a_{\epsilon\nu} = \delta_{\epsilon\nu}a_{\epsilon\nu} = a_{\nu\nu} \qquad (6.33)$$

The vector $\nabla_\mu^{(4)} = \partial/\partial x_\mu$ behaves as a four-vector. In order to make that plausible consider the scalar $x_\mu x_\mu$ (which is invariant) and let $\nabla^{(4)}$ work on it; it is seen that

$$\frac{\partial}{\partial x_\nu}(x_\mu x_\mu) = 2\delta_{\mu\nu}x_\mu = 2x_\nu \;\rightarrow\; L_{\epsilon\nu}\nabla_\nu^{(4)} = \nabla_\epsilon'^{(4)} \qquad (6.34)$$

An application of these transformation properties will be encountered in the section on electromagnetism, below.

6.6.2 Particle in a Constant Force Field

If a particle feels a constant force in its rest system and this force is given by $F_1 = (gm, 0, 0)$, then, from the spatial part of the Minkowski equation $dp_1/d\tau = mg$, it follows that $\gamma u_1 = gt$ and a further integration with respect to time yields

$$x(t) = \frac{c^2}{g}\left(\sqrt{1 + \frac{g^2 t^2}{c^2}} - 1\right) \qquad (6.35)$$

The relation $x(t)$ is the world line of the particle.

The classical result would obviously be $x = 1/2gt^2$. For long times, the velocity approaches the speed of light; naturally, it can never exceed it and as it also never entirely reaches it, the world line of the particle will be approximately the world line of a light ray. This is illustrated in Fig. 6.4. Note, however, that the momentum *can* increase without upper limit.

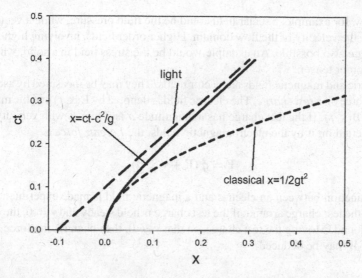

Fig. 6.4 World line of a particle under a constant force field (solid line), the classical result (short dash) and the world line of a light ray that is the limit of the particle world line (long dash)

The world line of the light ray depicted in Fig. 6.4 never meets the world line of the particle. A light ray that is emitted from $x(0) = A$, with $0 > A > c^2/g$ by contrast *will* meet the particle somewhere. A light ray emitted in the negative direction from $x(0) > 0$ will also encounter the particle. A traveller on the particle will thus be able to 'see' light that is emitted from locations at time $t = 0$ that are confined to $x(0) > c^2/g$. Light emitted from locations of $x(0)$ less than this will never be visible to the traveller. There is therefore an *event horizon*. Such a feature is typical for *accelerated systems*.

6.7 Special Relativity and Electromagnetic Theory

From everything that has been stated in the previous sections, it is clear that a treatment of Maxwell's equations is necessary to understand the context of relativity theory. Maxwell's equations describe electric and magnetic fields.

6.7.1 Fields and Potentials

A field is a physical property defined over the whole of space and at all times. Two types of fields are distinguished here: scalar fields and vector fields. The latter is then a vector prescription as a function of space and time, while the former gives one quantity. The use of fields is not only confined to electromagnetic properties. In fluid flow, for example, a scalar field could be the fluid pressure, while a vector field could be the velocity of the flow domain. Higher order fields, involving higher order tensors are also possible. An example would be the stress field in a solid, which is a second-order tensor.

Electric and magnetic fields are vector fields. They may be measured by ascertaining the force on a *test charge*. The electric field is denoted by $\mathbf{E}(\mathbf{x}, t)$ and the magnetic field by $\mathbf{B}(\mathbf{x}, t)$. If the test charge has a magnitude q and moves with velocity $\mathbf{v}$, the force exerted on it by the electromagnetic fields, the *Lorentz force* is

$$\mathbf{F} = q\,(\mathbf{E} + \mathbf{v} \times \mathbf{B}) \tag{6.36}$$

The distinction between an electric and a magnetic field is made experimentally by moving the test charge around; if the test charge is held steady and $\mathbf{v} = 0$, the electric field is found. Moving the test charge, so that $\mathbf{v} \neq 0$, then changes the force and the value of $\mathbf{B}$ may be deduced.

6.7.1.1 Units

A word on *electrical units* is necessary. Two systems are in general use.

The *SI system* (also called the *rationalised system*) uses the metre, kilogram, second and Ampère (mksA) as basic units, from which basis all other physical quantities can be expressed. So, for example, the Newton (N) to measure force is $1\,\mathrm{N} = 1\,\mathrm{kgms}^{-2}$, as follows from $F = ma$. In this system, the unit of electrical charge is the Coulomb (C)—$1C = 1As$. The electric field is measured in $N/C = kgm\,A^{-1}s^{-3}$, which is also known as a Volt per metre; $1\,\mathrm{Volt} = 1\,\mathrm{kgm}^2\mathrm{s}^{-3}\mathrm{A}^{-1}$. The magnetic field $\mathbf{B}$ (also known as the magnetic flux density or magnetic induction) has the unit $kg\,A^{-1}s^{-2}$, which is known as the *Tesla*, T. Another unit is the Weber (W) and $1\,\mathrm{W} = 1\,\mathrm{Tm}^2 = 1\,\mathrm{Vs}$.

The Gaussian unit system is based on the cm, gram, second set and a different unit of charge, the statCoulomb (statC). $1\mathrm{statC} = 1/\sqrt{4\pi\epsilon_0}C$, where ϵ_0 is the permittivity of the vacuum—$\epsilon_0 = 8.854 \times 10^{-12}\mathrm{A}^2\mathrm{s}^4\mathrm{kg}^{-1}\mathrm{m}^{-3}$. A different expression of the Lorentz force is necessary, using $\mathbf{v}/c$, rather than $\mathbf{v}$. Therefore, the Gaussian units are obtained from the SI units by inserting powers of 10 and factors of the speed of light c. The unit of the electric field in the Gaussian system is $cm^{-1/2}g^{1/2}s^{-1}$ and that is also the unit of the magnetic field.

In the text below, either SI units are preferred, though—in order to align with the prevailing literature—at times Gaussian units will be employed.

6.7.1.2 Potentials

Rather than specifying a vector field everywhere at all times, the curl and the divergence may be given, denoted by $\nabla\times$ and $\nabla\cdot$, respectively, see Sect. A.3 and Boas (1983). The div is associated with *sources* of the field; it is as if the field emanates from a certain source (or, negatively, disappears into a sink). The curl of a field is connected with—as the word implies—rotations or vortices in the field. A theorem states that any vector field $\mathbf{F}$ can be uniquely decomposed as

$$\mathbf{F} = \mathbf{F}_d + \mathbf{F}_c + \mathbf{F}_0 \text{ where } \nabla \times \mathbf{F}_d = 0, \ \nabla \cdot \mathbf{F}_c = 0 \text{ and } \mathbf{F}_0 \text{ is constant} \quad (6.37)$$

Because of their physical implications, specifying the operators curl and div of a field can be related to either their sources or their 'whirlyness', which makes the physics transparent. For example, an electric field may be associated with a point charge, as was found in the experiments by Coulomb.

Frequently, *potentials* may be employed. A scalar potential Φ gives the field as a gradient and is possible when the curl of the field vanishes. A vector potential $\mathbf{A}$, by contrast, produces the field as a curl and is possible when the div of a field vanishes. These properties follow mathematically from $\nabla \times \nabla \cdot \Phi = 0$ and $\nabla \cdot \nabla \times \mathbf{A} = 0$. Potentials are a really useful mathematical tool and it is seen here that the decomposition of a field in a div-free and curl-free part leads directly to their deployment.

6.7.2 Charges and Currents

Maxwell's equations give the curls and divs of electric and magnetic fields from the distribution of charges and currents. The latter is most easily described as densities; the charge density $\rho(\mathbf{x}, t)$ specifies the amount of electrical charge in a unit volume. The charge current density $\mathbf{j}(\mathbf{x}, t)$ denotes the charge density times its velocity. The actual current, that is the amount of charge per unit of time, that flows through a surface with area S and unit normal $\mathbf{n}$ is, (Fig. 6.5)

$$I = \iint_S \mathbf{j}(\mathbf{x}, t) \cdot \mathbf{n} \, dS \quad (6.38)$$

Fig. 6.5 Charges flowing through a surface with area S and unit normal $\mathbf{n}$

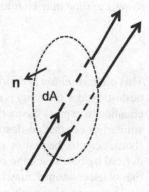

The total charge in a volume V is

$$Q(t) = \iiint_V \rho(\mathbf{x}, t)dV \tag{6.39}$$

The evolution of the charge distribution as the charges move is studied as follows. For a given volume V with area S (and unit outward normal $\mathbf{n}$), charges flowing in and out are captured by stating the current that goes through the surface area. For a given surface, this is given by Eq. (6.38) and involves the current density field

$$I_{\text{eff}} = \iint_S \mathbf{j}(\mathbf{x}, t) \cdot \mathbf{n}dS \tag{6.40}$$

If as much charge flows into the volume as flows out in a unit time, then $I_{\text{eff}} = 0$. If, however, some charge is accumulated or lost in the volume then I_{eff} represents the change in charge content in the volume Q_V per unit time and therefore

$$I_{\text{eff}} = \frac{\partial Q_V}{\partial t} = \frac{\partial}{\partial t} \iiint_V \rho(\mathbf{x}, t)dV \tag{6.41}$$

Evaluating the amount of charge that flows in and out of a volume V per unit of time is equal to the change in charge inside the volume

$$\frac{\partial}{\partial t} \iiint_V \rho(\mathbf{x}, t)dV = - \iint_A \mathbf{j} \cdot \mathbf{n}dS \tag{6.42}$$

Or, equivalently, using Gauss' theorem to turn the surface integral over a closed surface into a volume integral

$$\frac{\partial}{\partial t} \iiint_V \rho(\mathbf{x}, t)dV + \iiint_V \operatorname{div} \mathbf{j}(\mathbf{x}, t)dV = 0 \tag{6.43}$$

The volume is fixed in space and arbitrary, so the integration and differentiation with respect to time may be interchanged. It follows that

$$\frac{\partial \rho(\mathbf{x}, t)}{\partial t} + \operatorname{div} \mathbf{j}(\mathbf{x}, t) = 0 \tag{6.44}$$

This is the *equation of continuity*. Essentially, it expresses the fact that charge cannot be destroyed. Continuity is not just valid for charge. Fluid and gas flows obey a similar equation to express the fact that mass cannot be destroyed. Other physical quantities, similarly, obey the 'no-destruction-possible' paradigm; think of the motion of crowds (bodies cannot be lost or gained), money (the amount of money in an economy is defined by how much the central banks allow in and if any is lost then the effects of the 'black economy' may be quantified using a continuity equation).

The connection with relativity theory is now made. First, it is noted that the observed current density vector field depends on the velocity with which an observer travels. For example, if an observer travels with exactly the same velocity as the current flow, then the **j** vector is actually zero. The charge density is also subject to observational variation, as the volume is vulnerable to Lorentz contraction. Therefore, the charge and current densities transform in some way under a Lorentz transformation. The equation of continuity shows how this is done.

From (6.44), it follows that

$$\frac{\partial \rho(\mathbf{x}, t)}{\partial t} + \text{div } \mathbf{j}(\mathbf{x}, t) = 0 \rightarrow ic\frac{\partial \rho}{\partial (ict)} + \frac{\partial j_k}{\partial x_k} = 0 \qquad (6.45)$$

This suggests that

$$\begin{bmatrix} j_1 \\ j_2 \\ j_3 \\ ic\rho \end{bmatrix} \qquad (6.46)$$

is a four-vector. Call this vector j_μ, then the equation of continuity reads $\partial j_\mu/\partial x_\mu = 0$. Here is an example of the use of $\nabla^{(4)}$ and the fact that continuity is expressed as an inner product implies that it holds in all inertial systems. The universality of the equation of continuity ensures that the current-density/charge-density four-vector has a solid physical foundation.

The transformation of the current-charge-density four-vector is carried out with the aid of a Lorentz transformation: $j'_\mu = L_{\mu\nu} j_\nu$.

6.7.3 Maxwell's Equations in Vacuum

The equations put forward by Maxwell relate the electric and magnetic field strengths to the charge density and current density fields. They do so by stating what the divergence and curl of these fields are. There are four equations. The electric field is denoted by **E** and the magnetic field by **B**; the charge density is ρ and the current density **j**. All these quantities are functions of space and time (in SI units):

$$\nabla \cdot \mathbf{E} = \frac{\rho}{\epsilon_0} \qquad (6.47)$$

$$\nabla \cdot \mathbf{B} = 0 \qquad (6.48)$$

$$\nabla \times \mathbf{E} = -\frac{\partial \mathbf{B}}{\partial t} \qquad (6.49)$$

$$\nabla \times \mathbf{B} = \mu_0 \left(\mathbf{j} + \epsilon_0 \frac{\partial \mathbf{E}}{\partial t} \right) \qquad (6.50)$$

It is noted that these equations are valid in vacuum; ϵ_0 and μ_0 are the *permittivity* and the *permeability* of free space. These properties can be measured: $\epsilon_0 = 8.8541878128(13) \times 10^{-12}$ Fm^{-1} (Farad per metre; in SI base units $A^2 s^4 kg^{-1} m^{-3}$);

$\mu_0 = 4\pi \times 10^7 \, \text{Hm}^{-1}$ (Henry per metre; in SI base units $A^{-2}s^{-2}kgm$). Below, the significance of these constants will be discussed further. In addition to Maxwell's equations, the equation of continuity must hold

$$\frac{\partial \rho}{\partial t} + \nabla \cdot \mathbf{j} = 0 \tag{6.51}$$

It is a simple exercise to show that Maxwell's equations are compatible with the equation of continuity. The character of the equations is revealed by studying their solutions and verifying the outcome to experimental configurations. A standard work on this subject, with numerous examples, is Jackson (1962).

6.7.3.1 Static Electric Fields

A very simple case of a configuration is the configuration of a sphere of radius R which is homogeneously charged with a constant charge density ρ_0. There is no time-dependence and as the charges stand still, the current density is zero. The first of Maxwell's equations is integrated over a sphere of radius r enclosing a volume $V(r)$

$$\iiint_{V(r)} \nabla \cdot \mathbf{E} dV = \iiint_{V(r)} \frac{\rho}{\epsilon_0} dV \tag{6.52}$$

The problem possesses spherical symmetry, so it makes sense to change to polar coordinates on the right-hand side, while the integral on the left-hand side is transformed using Gauss' theorem with a unit outward vector $\mathbf{n}$

$$\iint_{S(r)} \mathbf{E} \cdot \mathbf{n} dS = \text{ if } r < R: \ \frac{4\pi}{3\epsilon_0} \rho_0 r^3 \text{ if } r \geq R: \ \frac{4\pi}{3\epsilon_0} \rho_0 R^3 \tag{6.53}$$

Again invoking spherical symmetry, the surface integral is done. The electric field is normal to the surface: $\mathbf{E} \cdot \mathbf{n} = E_n(r)$

$$4\pi r^2 E_n(r) = \text{ if } r < R: \ \frac{4\pi}{3\epsilon_0} \rho_0 r^3 \text{ if } r \geq R: \ \frac{4\pi}{3\epsilon_0} \rho_0 R^3 \tag{6.54}$$

It follows that

$$\text{if } r < R: \ E_n = \frac{\rho_0}{3\epsilon_0} r \text{ if } r \geq R: \ E_n = \frac{Q}{4\pi \epsilon_0 r^2} \tag{6.55}$$

where Q is the total charge in the sphere with radius R. The case $r \geq R$ represents *Coulomb's law*. This is a very special case of Maxwell's equations and one of the earliest systematic findings in electric field research. The outcome would be no different if the charge were not uniformly distributed in the volume of the sphere, but, for example, all concentrated on the surface of the sphere. The 'inverse square law' is a very general finding in physics where a field is caused by a source.

6.7.3.2 Static Magnetic Fields

A static magnetic field can be achieved as follows. Let a current flow through an infinite circular cylinder of radius a—a wire, essentially. The current has a constant density j_0 parallel to the axis while $r < a$ and zero outside it, where r is the radial coordinate from the centre of the cylinder. The fourth of Maxwell's equations can be used to give

$$\nabla \times \mathbf{B} = \mu_0 \mathbf{j} \tag{6.56}$$

Now consider a cylindrical surface of radius r and integrate over the cross-sectional surface with unit normal $\mathbf{n}$

$$\iint_{S(r)} (\nabla \times \mathbf{B}) \cdot \mathbf{n} dS = \mu_0 \iint_{S(r)} \mathbf{j} \cdot \mathbf{n} dS \tag{6.57}$$

Applying Stokes' theorem to the left-hand-side and doing the integral of the right-hand side yields

$$\int_{\lambda(r)} \mathbf{B} \cdot \mathbf{m} d\lambda = \text{if } r < a \ \pi r^2 \mu_0 j_0 \text{ if } r \geq a \ \mu_0 I \tag{6.58}$$

Here, $\lambda(r)$ is the circular loop around the cylinder and $\mathbf{m}$ the unit vector along the loop, while I is the total current through the cylinder. So, it is seen that $\mathbf{B}$ is constant along the loop and $\mathbf{B} \cdot \mathbf{m} = B_m$ is independent of the angle. The integral is elementary

$$\text{if } r < a : B_m = \frac{1}{2} r \mu_0 j_0 \text{ if } r \geq a: \ B_m = \mu_0 \frac{I}{2\pi r} \tag{6.59}$$

The latter result is a manifestation of *Ampère's law*, in words: 'The magnetic field created by an electric current is proportional to the size of that electric current with a constant of proportionality equal to the permeability of free space'. If the current is reversed the magnetic field also changes direction, though its magnitude remains the same.

For the geometry outlined here, there is also an electric field; it is left as an exercise for the reader to obtain it using the first of Maxwell's equations.

6.7.3.3 Maxwell's Equations in Terms of Potentials

One of the most intriguing features of Maxwell's equations in vacuum is the fact that the magnetic field is divergence-free. In other words, it has no sources, or, alternatively put, there are no *magnetic monopoles* in Nature. This is quite different for electric fields where *point charges* may be identified. Elementary particles, such as electrons, are obvious examples. Magnetic fields are entirely associated with closed-loop currents. There is no fundamental reason why magnetic monopoles should not exist; they have just never been observed. Dirac has shown that if magnetic monopoles exist then electric charge should be quantised. In fact, electric charge *is* quantised, but that does not necessarily mean that monopoles exist. There are 'quasi monopoles', not elementary particles, but solid-state objects that from afar would appear to have a 'magnetic charge'.

Here, the concern is with free space and $\nabla \times \mathbf{B} = 0$ is assumed to be true. That means that the magnetic field can be obtained from a vector potential $\mathbf{A}$

$$\mathbf{B} = \nabla \times \mathbf{A} \tag{6.60}$$

Substituting in the third of Maxwell's equation takes the form

$$\nabla \times \mathbf{E} = -\frac{\partial}{\partial t}(\nabla \times \mathbf{A}) = -\nabla \times \left(\frac{\partial \mathbf{A}}{\partial t}\right) \tag{6.61}$$

It follows that $\mathbf{E} + \partial \mathbf{A}/\partial t$ is curl-free, which implies that there is a scalar potential Φ, such that

$$\mathbf{E} + \frac{\partial \mathbf{A}}{\partial t} = -\nabla \Phi \tag{6.62}$$

Two of Maxwell's equations have now been used up; the other two read

$$\nabla \times (\nabla \times \mathbf{A}) = \mu_0 \left(\mathbf{j} + \epsilon_0 \frac{\partial}{\partial t}(\nabla \Phi) - \epsilon_0 \frac{\partial^2 \mathbf{A}}{\partial t^2}\right) \tag{6.63}$$

$$-\nabla^2 \Phi - \frac{\partial \nabla \cdot \mathbf{A}}{\partial t} = \frac{\rho}{\epsilon_0} \tag{6.64}$$

Using the identity $\nabla \times (\nabla \times \mathbf{A}) = \nabla(\nabla \cdot \mathbf{A}) - \nabla^2 \mathbf{A}$ these turn into the set

$$\nabla^2 \mathbf{A} - \mu_0 \epsilon_0 \frac{\partial^2 \mathbf{A}}{\partial t^2} = -\mu_0 \left(\mathbf{j} + \epsilon_0 \frac{\partial}{\partial t}(\nabla \Phi)\right) + \nabla(\nabla \cdot \mathbf{A}) \tag{6.65}$$

$$-\nabla^2 \Phi - \frac{\partial \nabla \cdot \mathbf{A}}{\partial t} = \frac{\rho}{\epsilon_0} \tag{6.66}$$

There is still a certain amount of freedom in the system in that $\nabla \cdot \mathbf{A}$ may still be chosen freely. This will be done in such a way that the resulting equations are Lorentz invariant. It is seen that one four-vector is already discernable: $[\mathbf{j}, ic\rho]$. The first equation contains three equations and the second equation is a single one. The following choice for $\nabla \cdot \mathbf{A}$ is made to effect a substantial simplification

$$\nabla \cdot \mathbf{A} = -\epsilon_0 \mu_0 \frac{\partial \Phi}{\partial t} \tag{6.67}$$

This choice is called the *Lorentz gauge*. Using it renders the equations in the following form:

$$\nabla^2 \mathbf{A} - \mu_0\epsilon_0 \frac{\partial^2 \mathbf{A}}{\partial t^2} = -\mu_0 \mathbf{j} \qquad (6.68)$$

$$\nabla^2 \Phi - \epsilon_0\mu_0 \frac{\partial^2 \Phi}{\partial t^2} = -\frac{\rho}{\epsilon_0} \qquad (6.69)$$

Below, it will be shown that the speed of light equals $c = 1/\sqrt{\epsilon_0\mu_0}$, or $\epsilon_0 = 1/(\mu_0 c^2)$. The form $\mathbf{x} \cdot \mathbf{x} - c^2 t^2$ is Lorentz-invariant, so it is seen that the potentials can be grouped into a four-vector

$$A_\mu \rightarrow \begin{bmatrix} A_1 \\ A_2 \\ A_3 \\ i\Phi/c \end{bmatrix} \qquad (6.70)$$

Use the notation $\Box = \nabla^{(4)} \rightarrow \nabla^2 - 1/c^2 \partial^2/\partial t^2$, or $\Box \rightarrow \partial^2/\partial x_\mu^2$ (a four-dimensional Laplacian), then the Maxwell equations take a very compact form, together with the Lorentz gauge

$$\Box A_\mu = -\mu_0 j_\mu \text{ with } \frac{\partial A_\mu}{\partial x_\mu} = 0 \qquad (6.71)$$

6.7.3.4 Light

Electromagnetic radiation is a solution of Maxwell's equations as a special case. The situation is as follows. Light is generated by an apparatus in which charges are oscillated. When the light has been created, the apparatus can be folded up and put away; however, that is not the end of the light. The radiation goes on existing and travels on into space, even though there are no charges or currents. That implies that a solution to Maxwell's equations may exist in the absence of charges and currents, that is with $j_\mu = 0$. Again, it is emphasised that solutions in free space are considered only.

The equation that needs to be solved is

$$\Box A_\mu = 0 \text{ with } \frac{\partial A_\mu}{\partial x_\mu} = 0 \qquad (6.72)$$

In order to create a solution, introduce a four-vector k_μ, such that $k_\mu k_\mu = 0$. Now, a solution is constructed as an arbitrary function of the inner product $k_\nu x_\nu$ of the form

$$A_\mu(x) = e_\mu f(k_\nu x_\nu) \qquad (6.73)$$

where the four components of e are chosen such that $e_\lambda k_\lambda = 0$. It is easily verified that this is a solution of Eq. (6.72), just substitute:

$$\frac{\partial A_\mu}{\partial x_\mu} = e_\mu f'\,(k_\nu x_\nu)\,k_\mu = 0 \text{ and } \Box A_\mu = e_\mu f''\,(k_\lambda x_\lambda)\,k_\nu k_\nu = 0 \qquad (6.74)$$

How to interpret this solution? As an example take a plane wave $A_\mu = e_\mu \sin(k_\lambda x_\lambda)$ and choose a coordinate frame in which $k = [k_1, 0, 0, ik_1]$ (this satisfies $k_\nu k_\nu = 0$). At time $ct = 0$, the wave has the form $e_\mu \sin(k_1 x_1)$, which has zeros in the locations $x_1 = 0, \pi/k_1, 2\pi/k_1 \dots$. At a later time, $x_4 = ict$, the form is $e_\mu \sin(k_1 x_1 + ik_1 x_4)$, which has zeros in $x_1 = ct, ct + \pi/k_1, ct + 2\pi/k_1 \dots$. The zeros have shifted by an amount ct; therefore, they have moved with the speed of light. *The wave propagates with the speed of light.* Also, the three-vector **k** dictates the direction of wave propagation. Note that this result is directly a consequence of the ansatz made in the previous section—$c = 1/\sqrt{\epsilon_0 \mu_0}$—which permitted the Maxwell equations to be written in Lorentz-invariant form.

For any given k, there are three independent choices that can be made for e. For example, in the case $[k_1, 0, 0, ik_1]$, these are

$$e_\mu^{(1)} = \begin{bmatrix} 0 \\ 1 \\ 0 \\ 0 \end{bmatrix} \quad e_\mu^{(2)} = \begin{bmatrix} 0 \\ 0 \\ 1 \\ 0 \end{bmatrix} \quad e_\mu^{(3)} = \begin{bmatrix} 1 \\ 0 \\ 0 \\ i \end{bmatrix} \qquad (6.75)$$

The vector $e_\mu^{(3)}$ is a multiple of k itself and it is easily shown that a function of the form $k_\mu f(k_\nu x_\nu)$ yields **E** and **B** fields that are zero. Therefore, $e_\mu^{(3)}$ does not correspond to a physically realistic situation. The two possibilities $e_\mu^{(1)}$ and $e_\mu^{(2)}$ correspond to the two possible *transversal polarisations* of light.

The four-vector k is called the *wave vector*. From the example of the solution $A_\mu = e_\mu \sin(k_\lambda x_\lambda)$, it is seen that the light propagates with a wavelength $\lambda = 2\pi/k_1$ and has frequency $f = c/\lambda$. The circular frequency is $\omega = 2\pi f = ck_1$. From the fact that $k_\mu k_\mu = 0$, it is deduced that k has the properties of a light-like four-vector. Indeed, transforming to another system with velocity v, it follows that $0 = k_\mu k_\mu = L(v)_{\mu\nu} k'_\nu L(v)_{\mu\epsilon} k'_\epsilon = L^{-1}(v)_{\nu\mu} k'_\nu L(v)_{\mu\epsilon} k'_\epsilon = \delta_{\nu\epsilon} k'_\nu k'_\epsilon = k'_\nu k'_\nu$. This suggests that the wave vector transforms as a four-vector. Experiments have been performed which show that that is actually true. Taking the wave vector as in the example above, that is $k_\mu = [k_1, 0, 0, ik_1]$, a Lorentz transformation with a velocity v in the same direction as **k** turns it into the form

$$\begin{bmatrix} k_1 \frac{1-v/c}{\sqrt{1-v^2/c^2}} \\ 0 \\ 0 \\ ik_1 \frac{1-v/c}{\sqrt{1-v^2/c^2}} \end{bmatrix} \qquad (6.76)$$

The wavelength observed in the moving frame has therefore changed from λ to λ', with

$$\frac{\lambda'}{\lambda} = \frac{\sqrt{1-v^2/c^2}}{1-v/c} \qquad (6.77)$$

For positive v the wavelength is thus greater—so-called *red shift*—and for negative v it is smaller—termed *blue shift*. So, using the Lorentz transformation the *Doppler effect* can be evaluated. The effect so ascertained is the longitudinal effect, which is produced by a velocity in the same direction as the vector $\mathbf{k}$. There is also a transversal effect, associated with a velocity normal to the direction of the wave vector. The latter is independent of the sign of v and of magnitude $\lambda'/\lambda = 1/\sqrt{1 - v^2/c^2}$.

The solution of the potential fields is employed to obtain the electric and magnetic fields that characterise light. As an example use the solution that has been put forward above $A_\mu = a_1 \sin(x_1 k_1 - ctk_1)[1, 0, 0, 0]$. The magnetic field follows from $\mathbf{B} = \nabla \times \mathbf{A}$, which results in $\mathbf{B} = a_1 k_1 \cos(x_1 k_1 - ctk_1)(0, 0, 1)$; the electric field is obtained from $\mathbf{E} = -\partial \mathbf{A}/\partial t$, which is $\mathbf{E} = a_1 ck_1 \cos(x_1 k_1 - ctk_1)(1, 0, 0)$. Now, it is observed that $\mathbf{B} \perp \mathbf{E}$. That is not just true for this particular example, but is a general property of light in a vacuum. It is also seen that electric and magnetic fields are in phase, which—again—is a general property for plane light waves.

It has been shown that a large number of phenomena that are explained from the basic properties of light can be deduced from the solutions of Maxwell's equations in the context of special relativity theory. Optical effects, such as interference, for instance, are convincingly explained by the wave character of light. But not all experiments on light can be accommodated. In particular, black body radiation, or the photo-electric effect, appeared not to follow from the solutions of Maxwell's equations. That was problematic and in 1901 (even before Einstein published his special relativity paper) Max Planck (1858–1947) came up with the revolutionary idea that light sometimes has the character of a stream of energetic particles, each with an energy of $E = hf$, where h is Planck's constant, which is an amazingly small number $h = 6.62607015 \times 10^{-34}\,\mathrm{kgm^2s^{-1}}$. Because it is so small its effects are manifest, not on the home-kitchen-and-garden scale of Newtonian mechanics, but on the atomic and sub-atomic scale.

References

Boas ML (1983) Mathematical methods in the physical sciences. Wiley, New York

Einstein A (1905a) Zur Elektrodynamik bewegter Körper. Annalen der Physik 17(10):891–921

Einstein A (1905b) Ist die Trägheit eines Körpers von seinem Energieinhalt abhängig? Annalen der Physik 18(13):639–641

Einstein A (1913) Entwurf einer verallgemeinerten Relativitätstheorie und einer Theorie der Gravitation. Zeitschrift für Mathematik und Physik 62:225–244

Einstein A (1918) Prinzipielles zur allgemeinen Relativitätstheorie. Annalen der Physik 55(4):241–244

Grossmann M (1913) Entwurf einer verallgemeinerten Relativitätstheorie und einer Theorie der Gravitation. Zeitschrift für Mathematik und Physik 62:245–261

Jackson JD (1962) Classical elctrodynamics. Wiley, New York

Koenders MAC (2020) The physics of the deformation of densely packed granular materials. World Scientific, Singapore

Lorentz HA (1904) Electromagnetic phenomena in a system moving with any velocity smaller than that of light. Proc R Neth Acad Arts Sci 6:809–831

Maxwell JC (1873) A treatise on electricity and magnetism. Nature 7(182):478–480
Reynolds O (1903) The sub-mechanics of the Universe. Royal Society of London, Cambridge
Rindler W (1966) Special relativity. Oliver and Boyt, Edinburgh
Weinberg S (1972) Gravitation and cosmology. Wiley, New York

Chapter 7
Many-Particle Systems

Abstract Statistical methods are introduced to describe the collective behaviour of many-particle systems. Boltzmann's transport equation is introduced and, in the steady state, an equilibrium solution is found for the momentum distribution of a gas with an elastic collision rheology. The same distribution is obtained by requiring it to be the most probable one. The flow in phase space is discussed for which Liouville's theorem is relevant and then the H-theorem is derived as a consequence of the existence of a Lyapunov function. Thermodynamic quantities are introduced and their link with the equilibrium momentum and position distribution is explained. The properties of solids are briefly touched on and then the van der Waals gas is discussed. This leads to an exposition of the critical point, phase transitions, supersaturated gas and superevaporated liquid. The Master equation is introduced. Again, it is shown that there is a Lyapunov function, which leads to the H-theorem. The van Kampen expansion of the Master equation and its link with macroscopic systems is analysed. The linear noise approximation (the Fokker-Planck equation) is put forward; the diffusion approximation is treated. Examples of the diffusion equation and the heat equation are solved.

7.1 Introduction

Newtonian mechanics is very successful at describing the behaviour of a small number of mass points; a small number here is understood to be one, two or three. Computer simulations can do larger numbers of particles, of course, but their number is still limited. When the intention is to describe systems of particles that contain a very large number, such as a cubic centimetre of gas, Newtonian mechanics, in the form in which it has been presented so far, is clearly not practical. It is a useful exercise to calculate how many particles there are in a cubic centimetre of a gas, such as nitrogen, at room temperature at atmospheric pressure. The specific mass of air is 1.225 kg/m^3 and the mass of one nitrogen molecule is 4.65×10^{-26} kg, so the number of particles in a cubic centimetre is some 2.6×10^{19}. In a three-dimensional simulation, merely addressing the position vectors of this number, requires some 2×10^{11} Gbyte—assuming three bytes per number stored in memory. It is completely unrealistic to try to simulate such a system. The question is then whether it is possible to approach the problem in a way different from brute-force calculation and use *statistical methods* to make progress.

Setting up a suitable statistical methodology begins with the introduction of *phase space*. If a six-dimensional space is created in which the coordinates are the position **r** and the momentum **p**, each particle at some time point t will be represented by a point $(\mathbf{r}_\nu, \mathbf{p}_\nu)$. For N particles in the system $\nu = 1..N$; N is the very large number of the order of 10^{20}, say. All the particles in phase space will be represented by a cloud of points. The cloud will in general not have a uniform density, but it is possible to partition the phase space in small six-dimensional cubes of size $d\mathbf{r} \times d\mathbf{p}$. These cells are of so-called *physically infinitesimal* size, meaning there are many particles (phase points) in each of them, but they are still so small that differential calculus can be applied to them. The number in each cell is $f(\mathbf{r}, \mathbf{p}, t)d^3r d^3p$, so that f is a number density. The total number of particles is

$$N(t) = \int_V \int f(\mathbf{r}, \mathbf{p}, t)d^3r d^3p \qquad (7.1)$$

where V is the volume of the system (which confines the spatial integral). The integral over the momenta is taken to be infinite. The time plays a peculiar role here. In a closed system, no particles can get in or out, so N would be constant. That means that strictly speaking a time average also needs to be carried out. The study of many-particle systems is usually called *statistical mechanics* (or sometimes *kinetic theory*), Landau and Lifshitz (1976), Mandl (1988), Pendlebury (1985), Pointon (1967), Ter Haar (1966). Many books, all with their own emphasis, have been written on the subject and on the associated field of scientific endeavour of *thermodynamics*, Fermi (1956), Jui Sheng (1957), Saggion et al. (2019).

7.2 Boltzmann's Transport Equation

A system of particles is considered in which all have the same mass m and make-up. The word 'identical' might have been used, but in this subject that word has a rather special meaning when quantum mechanical processes are considered that involve indistinguishable particles. The particles in this section could be called 'classically identical', meaning they are identical, but distinguishable. They could—in principle—be tracked individually, though in an actual experiment that may be quite hard to do without disturbing the system to an unacceptable degree.

Having introduced phase space and its distribution function $f(\mathbf{r}, \mathbf{p}, t)$, the question is how f changes over time. At time t, the number of phase points in a cell is $f(\mathbf{r}, \mathbf{p}, t)d^3r d^3p$; at a time $t + \Delta t$, there are $f(\mathbf{r}, \mathbf{p}, t + \Delta t)d^3r d^3p$. The difference between these two is equal to

$$\frac{\partial f}{\partial t}\Delta t d^3r d^3p \qquad (7.2)$$

There are two reasons why $\partial f/\partial t$ may have a value. First, particles may leave the cell at $(\mathbf{r}, \mathbf{p})$ in a time interval Δt and end up in surrounding ones. Second, particles from surrounding cells may enter the cell at $(\mathbf{r}, \mathbf{p})$. There are two mechanisms by which these entries and departures may happen: natural drift and collisions between particles, in other words

$$\frac{\partial f}{\partial t} = \left(\frac{\partial f}{\partial t}\right)_{\text{drift}} + \left(\frac{\partial f}{\partial t}\right)_{\text{coll}} \tag{7.3}$$

These two mechanisms work independently of each other. Natural drift is described by an equation of continuity, which is valid, because no phase points are destroyed. Now, the equation of continuity needs to be considered in six dimensions

$$-\left(\frac{\partial f}{\partial t}\right)_{\text{drift}} = \frac{\partial f}{\partial \mathbf{r}}\dot{\mathbf{r}} + \frac{\partial f}{\partial \mathbf{p}}\dot{\mathbf{p}} \tag{7.4}$$

The time derivatives are $\dot{\mathbf{r}} = p/m$ and $\dot{\mathbf{p}} = \mathbf{F}$, where $\mathbf{F}$ is an external force—gravity, for example.

For the rate of change in f due to collisions, the introduction of a transition probability is necessary. Collisions could take place anywhere in the system and the actual collision process depends on the momenta of the participating particles only. The number of particles that take part in an event is important. Here, *binary collisions* are considered in which two particles participate. (The theory can be modified to allow for collisions in which simultaneous encounters involving more than two particles can be described.) The transition probability depends on two particles with momenta $\mathbf{q}$ and $\mathbf{q}'$ colliding and change the momenta to $\mathbf{p}$ and $\mathbf{p}'$. The transition probability per unit time is denoted by $W(\mathbf{q}, \mathbf{q}' \to \mathbf{p}, \mathbf{p}')$. The increase in the number of phase points in the cell at $\mathbf{p}$ due to those participating from cells at $\mathbf{q}$ and $\mathbf{q}'$ is

$$\int d^3 p' \left[W(\mathbf{q}, \mathbf{q}' \to \mathbf{p}, \mathbf{p}') f(\mathbf{q}) d^3 q f(\mathbf{q}') d^3 q' \right] \tag{7.5}$$

Summing over $\mathbf{p}'$ has been done, because all collisions that end up in $\mathbf{p}$ contribute. The decrease due to collisions out of the cell at $\mathbf{p}$ is

$$\int d^3 q \int d^3 q' \left[W(\mathbf{p}, \mathbf{p}' \to \mathbf{q}, \mathbf{q}') f(\mathbf{p}) d^3 p f(\mathbf{p}') d^3 p' \right] \tag{7.6}$$

Summing over all possible collisions yields the total change in f

$$\left(\frac{\partial f}{\partial t}\right)_{\text{coll}} = -f(\mathbf{p}) \int d^3 q \int d^3 q' \int W(\mathbf{q}, \mathbf{q}' \to \mathbf{p}, \mathbf{p}') f(\mathbf{p}') d^3 p' +$$
$$+ \int d^3 p' \int \int W(\mathbf{p}, \mathbf{p}' \to \mathbf{q}, \mathbf{q}') f(\mathbf{q}) f(\mathbf{q}') d^3 q d^3 q' \tag{7.7}$$

A further assumption is that the collisions are central-symmetric, in which case it can be shown that

$$W(\mathbf{q}, \mathbf{q}' \to \mathbf{p}, \mathbf{p}') = W(\mathbf{p}, \mathbf{p}' \to \mathbf{q}, \mathbf{q}') \tag{7.8}$$

Furthermore, it is assumed that the distribution does not depend on the position $\mathbf{r}$. Then the distribution function f satisfies the non-linear autonomous differential equation

$$\frac{\partial f}{\partial t} = \int d^3 p' \int \int W(\mathbf{p}, \mathbf{p}' \to \mathbf{q}, \mathbf{q}') \left[f(\mathbf{q}) f(\mathbf{q}') - f(\mathbf{p}) f(\mathbf{p}') \right] d^3 q d^3 q' \tag{7.9}$$

This is *Boltzmann's transport equation.*

The transition probability W represents the collision rheology. For spherically symmetric molecules an elastic collision may be assumed. That means that W is zero unless

$$\mathbf{q} + \mathbf{q}' = \mathbf{p} + \mathbf{p}' \text{ and } \frac{1}{2}(q^2 + q'^2) = \frac{1}{2}(p^2 + p'^2) \tag{7.10}$$

A steady equilibrium state is achieved when $\partial f / \partial t$ in Eq. (7.9) vanishes. A solution for f_{eq} can then only depend on the energy and the momentum, a solution that is known as the Maxwellian velocity distribution

$$f_{eq}(\mathbf{p}, \mathbf{x}) = \exp\left(\alpha + \zeta p^2 + \mathbf{k}.\mathbf{p}\right) \theta(\mathbf{x} \in V) \tag{7.11}$$

The θ function equals unity when $\mathbf{x}$ is inside the volume V and zero when $\mathbf{x}$ is outside it. Here, the position dependence of the distribution function is a simple cut-off at the volume boundaries. The constants α and ζ can be determined by requiring the total number of particles in the system to be N and the internal energy to be equal to a prescribed value U. The constant $\mathbf{k}$ only plays a role when the system flies by at a constant velocity and this may be ignored for a study of systems that stand still.

7.3 Maxwellian Obtained from Maximum Probability

There is an alternative method of deriving Boltzmann's velocity distribution f_{eq} and this method shines a very interesting light on the ideas of statistical mechanics. To that end phase, space is sub-divided into n small cells; these are numbered $i = 1...n$. The number of particles in each cell is n_i. There is no way of knowing which particle will occupy a certain cell at a given time. Over a long time, however, it may be assumed that each particle 'samples' the whole of phase space. The probability of finding n_i particles in a certain cell at a given time point is then a sensible concept, even though individual particles move around a great deal, in equilibrium—and again emphasising that the number of particles in each cell is very large—this number will be constant. The particles may be distributed over the cells. A key assumption is made: *a priori the probability of finding a certain number of particles in a cell is proportional to the volume of the cell.* The cells are chosen to have the same size, so this assumption implies that no weights have to be introduced for different cells. The number of ways N particles can be distributed such that there are n_1 in cell number 1, n_2 in cell number 2, n_3 in cell number 3, etc., is

$$W(n_1, n_2, n_3, ...) = \frac{N!}{n_1! n_2! n_3! ....} \tag{7.12}$$

Now the idea is that the most probable distribution is the one for which W is maximal under the condition that the total energy of the system is E and that the total number of particles is N. The question may be asked whether these are the only requirements that should be imposed and below more will be said about that. Instead of maximising W, its natural logarithm is maximised and Stirling's formula for the factorials is employed (these are after all large numbers): $N! \to N^N$. The functional that needs to be considered then is

$$\ln W(n_1, n_2, n_3, ...) + \alpha \left(\sum_i n_i - N \right) + \beta \left(\sum_i n_i H_i - E \right) \tag{7.13}$$

where α and β are Lagrange multipliers, H_i is the Hamiltonian in cell number i and

$$\ln W(n_1, n_2, n_3, ...) \to N \ln N - \sum_i n_i \ln n_i \tag{7.14}$$

Differentiating with respect to the 'variables' of the problem n_i and setting to zero gives

$$n_i = e^{-\alpha - 1} e^{-\beta H_i} \tag{7.15}$$

For free particles, the Hamiltonian is $p^2/(2m)$. Thus, it follows that the distribution found from the imposition of maximum probability is the same as the one obtained directly from Boltzmann's transport equation with a collision rheology as described. The distinguishing aspect of the method described in this section is that it does not depend on a collision rheology, so its applicability is much more general than the method discussed in Sect. 7.2.

7.4 Liouville's Theorem

The motion of the system in phase space is—on a microscopic level—ruled by Hamilton's equations. This will now be employed to come to more insight into the behaviour of the density function. A variation in $f(\mathbf{p}_i, \mathbf{r}_i)$ is represented as

$$df = \frac{\partial f}{\partial t} dt + \sum_i \frac{\partial f}{\partial \mathbf{p}_i} d\mathbf{p}_i + \sum_i \frac{\partial f}{\partial \mathbf{r}_i} d\mathbf{r}_i \tag{7.16}$$

Evaluating the variation along a line that goes through the phase points, in other words when the variations in momentum and position are such that $d\mathbf{p}_i = \dot{\mathbf{p}}dt$ and $d\mathbf{r}_i = \dot{\mathbf{r}}dt$, renders

$$\frac{df}{dt} = \frac{\partial f}{\partial t} + \sum_i \frac{\partial f}{\partial \mathbf{p}_i}\dot{\mathbf{p}}_i + \sum_i \frac{\partial f}{\partial \mathbf{r}_i}\dot{\mathbf{r}}_i \qquad (7.17)$$

Alternatively, the motion of the phase points can be regarded as a flow-through phase space. No phase points can be destroyed and therefore the equation of continuity in phase space is valid

$$\frac{\partial f}{\partial t} + \sum_i \left(\frac{\partial(f\dot{\mathbf{p}}_i)}{\partial \mathbf{p}_i} + \frac{\partial(f\dot{\mathbf{r}}_i)}{\partial \mathbf{r}_i} \right) = 0 \qquad (7.18)$$

Expanding the brackets yields

$$\frac{\partial f}{\partial t} + \sum_i \left(\frac{\partial f}{\partial \mathbf{p}_i}\dot{\mathbf{p}}_i + \frac{\partial f}{\partial \mathbf{r}_i}\dot{\mathbf{r}}_i \right) + f \sum_i \left(\frac{\partial \dot{\mathbf{p}}_i}{\partial \mathbf{p}_i} + \frac{\partial \dot{\mathbf{r}}_i}{\partial \mathbf{r}_i} \right) = 0 \qquad (7.19)$$

By using Hamilton's equations, the second term on the left-hand side is evaluated

$$\sum_i \left(\frac{\partial \dot{\mathbf{p}}_i}{\partial \mathbf{p}_i} + \frac{\partial \dot{\mathbf{r}}_i}{\partial \mathbf{r}_i} \right) = \sum_i \left(-\frac{\partial^2 H}{\partial \mathbf{p}_i \partial \mathbf{r}_i} + \frac{\partial^2 H}{\partial \mathbf{r}_i \partial \mathbf{p}_i} \right) = 0 \qquad (7.20)$$

And therefore,

$$\frac{df}{dt} = 0 \text{ and } \frac{\partial f}{\partial t} = -\{f, H\} \qquad (7.21)$$

It follows that the flow in phase space behaves as an incompressible fluid; this fact is known as *Liouville's theorem*.

Returning now to the question from the previous section whether the imposition of the total energy and the total number of particles is sufficient, or whether there are other quantities that need to be taken into account. Essentially what is proposed is that the functional dependence of the distribution function $f(\mathbf{p}_i, \mathbf{r}_i)$ is restricted to a dependence of the form $f[H(\mathbf{p}_i, \mathbf{r}_i)]$. For the function to be single-valued and continuous, it can depend only on the 'uniform' integrals of motion, by which is meant integrals of motion that are single-valued continuous functions of the momenta and position. The energy is obviously a prime candidate. However, it is possible that something like a component of the angular momentum may be a relevant integral too. For example, if a gas is enclosed in a circular cylinder with perfectly smooth walls, then the component of the angular momentum about the axis of the cylinder is an integral of motion, for the forces associated with the walls (the external forces) in a collision preserve this integral. However, if the wall is even ever so slightly imperfect, then the integral will be destroyed as there is no cylindrical symmetry any more. Therefore, the assumption is that in realistic systems the energy is the only uniform integral and the distribution function in equilibrium, which is the most probable distribution function, depends on the Hamiltonian only. Liouville's theorem reinforces this concept, as the rate of change of the distribution function is obtained from the Poisson bracket with the Hamiltonian, as shown in Eq. (7.21).

7.5   The H-Theorem

Equation (7.9) possesses a function of Lyapunov, which gives information on the stability of the system. For simplicity, the position dependence will be ignored (f may be integrated over the volume and then divided by V), so f will be the velocity distribution. The Lyapunov function is called H (*not* the Hamiltonian) and is defined as

$$H = \int d^3p f(\mathbf{p}) \log \left[\frac{f(\mathbf{p})}{f_{eq}} \right] \tag{7.22}$$

To demonstrate that this is indeed a function of Lyapunov it has to be shown that $dH/dt \leq 0$, while the equal sign is valid when $f = f_{eq}$. The time derivative is easily obtained

$$\frac{dH}{dt} = \int d^3p \frac{\partial f}{\partial t} \left(\log \left(\frac{f}{f_{eq}} \right) + 1 \right) \tag{7.23}$$

For $\partial f / \partial t$, Boltzmann's transport equation is employed

$$\frac{dH}{dt} = \int d^3p \int d^3p' \int d^3q \int d^3q' W(\mathbf{p}, \mathbf{p}' \to \mathbf{q}, \mathbf{q}') \times$$
$$\times [f(\mathbf{q})f(\mathbf{q}') - f(\mathbf{p})f(\mathbf{p}')] \left(\log \left(\frac{f}{f_{eq}} \right) + 1 \right) \tag{7.24}$$

Now, this equation does not change when the integration variables $\mathbf{p}$ and $\mathbf{p}'$ are interchanged. Equally, the same follows when $\mathbf{q}$ and $\mathbf{q}'$ are swapped. Finally, when the pairs $(\mathbf{p}, \mathbf{p}')$ and $(\mathbf{q}, \mathbf{q}')$ change places the equation remains invariant as well. So, the four versions of Eq. (7.24) can be added together to give

$$4\frac{dH}{dt} = \int \int \int \int d^3p d^3p' d^3q d^3q' W(\mathbf{p}, \mathbf{p}' \to \mathbf{q}, \mathbf{q}') \times$$
$$\times [f(\mathbf{q})f(\mathbf{q}') - f(\mathbf{p})f(\mathbf{p}')] \log \frac{f(\mathbf{p})f(\mathbf{p}')}{f(\mathbf{q})f(\mathbf{q}')} \tag{7.25}$$

The expression on the second line is always negative and only zero when $f(\mathbf{q})f(\mathbf{q}') = f(\mathbf{p})f(\mathbf{p}')$. The latter means that the equilibrium distribution is returned.

W is always positive or zero and therefore it always holds that $dH/dt \leq 0$ and it is observed that H always decreases until the Maxwellian equilibrium solution is achieved. This analysis is known as the *H-theorem*. Its validity is rather more general than the one derived here for strictly collisional interactions, as will be further investigated below.

7.6 Thermodynamic Variables

Once the distribution, Eq. (7.15), is known macroscopic variables can be calculated. These variables can be expressed in two ways: either by specifying the quantities for a system of N particles, or by stating what the average value is *per particle*. For many macroscopic variables, the former can be obtained from the latter by simply multiplying by N. In that case, enlarging the system by a factor will simply increase the quantity by that factor. More formally, a process can be imagined by which both the number of particles N and the volume of the system V grow larger in such a way that the density $\rho = N/V$ remains constant (the temperature is also kept constant). The simultaneous limits $N \rightarrow \infty$, $V \rightarrow \infty$, N/V constant is known as the *thermodynamic limit*. Quantities that increase in this imaginary process that are proportional to N are called *extensive quantities*, while quantities that remain unaltered are referred to as *intensive quantities*. Quantities for the whole system and quantities per particle will be distinguished by supplying the latter with a tilde.

The first obvious quantity to evaluate is the *internal energy per particle* $\tilde{U}$, which follows from

$$\tilde{U} = \frac{\int_V d^3r \int d^3p\, H(\mathbf{p}, \mathbf{r}) n(\mathbf{p}, \mathbf{r})}{\int_V d^3r \int d^3p\, n(\mathbf{p}, \mathbf{r})} = \frac{\int_V d^3r \int d^3p\, H(\mathbf{p}, \mathbf{r}) e^{-\beta H(\mathbf{p}, \mathbf{r})}}{\int_V d^3r \int d^3p\, e^{-\beta H(\mathbf{p}, \mathbf{r})}} \qquad (7.26)$$

The denominator of this expression—but for a dimensional factor $\tilde{K}$—goes under the name *single particle partition function*, $\tilde{Z}$; it is immediately seen that

$$\tilde{U} = -\frac{\partial \ln \tilde{Z}}{\partial \beta} \qquad (7.27)$$

For a perfect gas, the Hamiltonian is $p^2/(2m)$ and the integral is easily done: $\tilde{Z} = \tilde{K}V(2\pi m/\beta)^{(3/2)}$; the internal energy is then $\tilde{U} = 3/(2\beta)$. From experiments, it is known that the heat capacity per particle at constant volume $(\partial \tilde{U}/\partial T)_V$ is $3k_B/2$ and—for an ideal gas—the constant β may be identified with the absolute temperature T and Boltzmann's constant k_B as $k_B T = \beta^{-1}$; $k_B = 1.380649 \times 10^{-23}$ JK^{-1}.

An ideal gas thermometer may at any time be used to measure the temperature; connecting it to any system leads to the insight that β is *always* related to the absolute temperature as $k_B T = \beta^{-1}$. The temperature of any system is in this way defined through the statistical behaviour of a perfect gas.

The quantity $\tilde{F}$ is defined from the partition function as

$$\tilde{Z} = e^{-\beta \tilde{F}} \quad \text{or} \quad \tilde{F} = -\frac{\ln \tilde{Z}}{\beta} \qquad (7.28)$$

For this equation to make sense $\tilde{Z}$ has to be non-dimensional and the factor $\tilde{K}$ must have the dimension of (length $\times$ momentum)$^{-3}$. It is emphasised again that the current approach is concerned with one particle, hence all the tildes. The meaning of $\tilde{F}$ can be found as follows.

Consider a system in a box of size V. To change this by an amount δV, a pressure p has to be exerted that corresponds to an amount of work done $\delta \epsilon$ equal to

$$- \delta \epsilon = p \delta V \tag{7.29}$$

Per particle this becomes

$$- \delta \tilde{\epsilon} = p \delta \tilde{V} \tag{7.30}$$

The minus sign is introduced to indicate that the work is done *by* the system. The partition function is

$$\tilde{Z} = e^{-\beta \tilde{F}} = \tilde{K} \int e^{-\beta \tilde{\epsilon}} d\omega \tag{7.31}$$

here ω stands for an infinitesimal of a phase space volume. The natural parameters here are the temperature (via β) and the energy $\tilde{\epsilon}$. Varying $1/\tilde{K}$ works out as

$$\delta \left(\frac{1}{\tilde{K}} \right) = 0 = \int \delta(\beta \tilde{F}) e^{\beta(\tilde{F} - \tilde{\epsilon})} d\omega - \int \tilde{\epsilon} \delta \beta e^{\beta(\tilde{F} - \tilde{\epsilon})} d\omega - \int \beta \delta \tilde{\epsilon} e^{\beta(\tilde{F} - \tilde{\epsilon})} \tag{7.32}$$

Using (7.30) then gives

$$\delta(\beta \tilde{F}) = \bar{\tilde{\epsilon}} \delta \beta - \beta p \delta \tilde{V} \tag{7.33}$$

where $\bar{\tilde{\epsilon}}$ is the mean internal energy per particle, which is essentially $\tilde{U}$. Now, introducing the heat increment $đ\tilde{Q}$ and the entropy $\tilde{S}$, the first and second laws of thermodynamics are deployed (the notation $đ$ implies that the variation is not a total one)

$$đ\tilde{Q} = \delta \tilde{U} + đ\tilde{A} \; ; \quad \delta \tilde{S} = \frac{đ\tilde{Q}}{T} \tag{7.34}$$

It then follows that

$$\delta(\beta \tilde{F}) = \bar{\tilde{\epsilon}} \delta \beta - \beta p d \tilde{V} = \delta \left(\frac{\tilde{U}}{k_B T} - \frac{\tilde{S}}{k_B} \right) \tag{7.35}$$

Therefore, apart from a possible constant, one may write

$$\frac{\tilde{F}}{T} = \frac{\tilde{U}}{T} - \tilde{S} \quad \text{or} \quad \tilde{F} = \tilde{U} - T\tilde{S} \tag{7.36}$$

The quantity $\tilde{F}$ is known as the *free energy per particle* and also that the pressure p equals $-\partial \tilde{F}/\partial \tilde{V}$. It is seen that the pressure is an intensive quantity, while the free energy (as all energy measures) is an extensive one. The entropy is an extensive quantity.

All these quantities are defined on a 'per particle' basis. To get the total quantities they must be multiplied by the number of particles N. For a perfect gas, for example, the equation of state is the well-known 'Boyle Gay-Lussac' law $p = k_B N T / V$.

A consistent expression for the partition function Z_N for N particles is necessary. From $\tilde{U} = -\partial\tilde{Z}/\partial\beta$, it is suggested that $Z_N = K_N \tilde{Z}^N$. Insight into the coefficient K_N can be gained by considering a perfect gas. The free energy is

$$F = -\frac{Z_N}{\beta} = -\frac{1}{\beta}\left[\ln K_N + \frac{3}{2}N\ln\left(\frac{2\pi m}{\beta}\right) + N\ln V\right] \qquad (7.37)$$

Choosing

$$K_N = \frac{\tilde{K}^N}{N!} \rightarrow \text{Stirling} \rightarrow \left(\frac{e}{N}\right)^N \tilde{K}^N \qquad (7.38)$$

Then

$$F = -\frac{1}{\beta}\left[N + N\ln\tilde{K} + \frac{3}{2}N\ln\left(\frac{2\pi m}{\beta}\right) + N\ln\frac{V}{N}\right] \qquad (7.39)$$

which satisfies the requirement

$$F = N\tilde{F} \qquad (7.40)$$

Now, it was seen that the dimension of $\tilde{K}$ is (length × momentum)$^{-3}$, which implies the dependence of an *action*. The parameter in physics that has the natural dimension of an action is Planck's constant h. While at this point no quantum mechanical aspects are expected, the usual choice for $\tilde{K}$ is indeed $\tilde{K} = h^{-s}$, where s is the number of degrees of freedom per molecule. For spherical particles with no internal degrees of freedom, such as those in a perfect gas, $s = 3$. For particles that have direction, such as diatomic molecules, the rotational degrees of freedom have to be added to the phase space and $s > 3$.

Boltzmann's H-theorem is now investigated in terms of thermodynamic variables. Recall that

$$f = \frac{e^{-\beta\epsilon}}{Z} \text{ or } f = KNe^{\beta(F-\epsilon)} \qquad (7.41)$$

Now,

$$H = \int f\ln f\,d\omega = \beta(F - \bar\epsilon) + \text{cnst} = -\frac{S}{k_B} + \text{cnst} \qquad (7.42)$$

It follows that for a closed system that evolves so that H always decreases, the entropy always increases, until a thermodynamic equilibrium state is reached.

For a link between microscopic assumptions and macroscopic, measurable, quantities of many-particle systems statistical mechanics is essential. The macroscopic quantities are thermodynamic ones; they *appear* to be single-valued parameters, though in fact, they are averages of a large number of fluctuating properties.

7.7   Other Phases

So far, the perfect gas was analysed only. In this section, statistical models of other phase states will be discussed. Crystalline solids are fairly easily captured in the framework of statistical mechanics. The description of fluids poses major problems and gross approximations are required to deal with these states of matter. All manner of other interesting phenomena, such as phase transitions and plasma states, are as yet not fully explored.

7.7.1   Crystallline Solids

A model of a crystal solid is easily made by trapping the atoms in a harmonic oscillator potential. Each atom then has a Hamiltonian

$$H(\mathbf{p}, \mathbf{q}) = \frac{p^2}{2m} + \frac{1}{2}kq^2 \tag{7.43}$$

Each atom is permitted to have the whole crystal volume at its disposal. The partition function $\tilde{Z}$ acquires an extra term compared with the perfect gas

$$\tilde{Z} = \tilde{K} \left(\frac{2\pi}{k\beta}\right)^{(3/2)} \left(\frac{2\pi m}{\beta}\right)^{(3/2)} \tag{7.44}$$

The internal energy per particle and the $\tilde{C}_V = \partial \tilde{U}_1 / \partial T$ are now easily obtained: $\tilde{C}_V = 3k_B$, which is indeed measured at high temperatures. At lower temperatures, quantum corrections come into play.

7.7.2   The Van Der Waals Gas

The physicist van der Waals attempted to describe a liquid state. In this state, the molecules have some liberty to move around, like in a gas, but they are also somewhat bound together. It is difficult to include such features in one model. A proposal for the interactive potential between two molecules has been made, the so-called *Lennard-Jones* potential. It is assumed that the molecules are spherical and that there is some critical distance r_0 (equivalent to the diameter of the molecules) between their surfaces, such that a slightly attractive potential kicks in when the distance r is less than $2r_0$. When $r < r_0$, there is a strong repulsive force, which describes the impenetrability of the molecules (not completely hard spheres, but pretty well so). The mathematical form of the potential is

$$\phi_{LJ} = 4\phi_0 \left[\left(\frac{r_0}{r}\right)^{12} - \left(\frac{r_0}{r}\right)^{6} \right] \tag{7.45}$$

Fig. 7.1 The Lennard-Jones potential

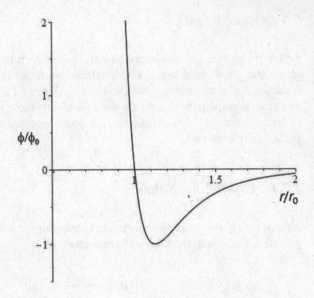

which has been plotted in Fig. 7.1. The value $r = r_0$ corresponds to $\phi_{\text{LJ}} = 0$ and the minimum value of ϕ_{LJ} is $\phi_{\text{LJ}}/\phi_{\text{min}} = -1$ at $r_{\text{min}}/r_0 = 21/6 = 1.122$. It is assumed that the intermolecular distance cannot be smaller than a molecular diameter (hard-particle repulsion) and two molecules cannot overlap. Therefore, the minimum inter-molecular distance from centre to centre is equal to the diameter d of one molecule. In terms of the Lennard-Jones potential set $d = r_{\text{min}}$, rather than $d = r_0$, because the Lennard-Jones interaction between two molecules is repulsive out to a separation of r_{min} as shown in Fig. 7.1.

In the van der Waals theory, the volume of a molecule ('excluded volume') is denoted by a variable b, so the free volume available for the molecules to move in is $V - Nb$, where $b = d^3 = r_{\text{min}}^3 = \sqrt{2}r_0^3$.

For $r > d$, the force between the gas molecules is assumed to be attractive, and the strength of the attraction depends on the distance between the molecules. In terms of the Lennard-Jones potential, this occurs for $r > r_{\text{min}}$ according to Fig. 7.1. This attractive part of the interaction is accounted for in an average way as follows, which is a 'mean-field' approximation where one ignores local fluctuations in the number density of molecules and short-range correlations between their positions. The number density of the molecules is N/V. The number dN of molecules that are at a distance between r and $r + dr$ from the central molecule is $dN = (N/V)dV$, where an increment of volume a distance r from the centre of the central molecule is $dV = 4\pi r^2 dr$. Thus, the total average attractive potential energy summed over these molecules, ϕ_{ave}, is

$$\phi_{\text{ave}} = \left(\frac{N}{V}\right)\frac{1}{2}\int_{r_{\text{min}}}^{\infty}\phi(r)dV = \left(\frac{N}{V}\right)\frac{4\pi}{2}\int_{r_{\text{min}}}^{\infty}\phi(r)r^2 dr \qquad (7.46)$$

where the prefactor of $1/2$ arises because the potential energy of interaction between a molecule and a neighbouring molecule is shared equally between them. In the van der Waals theory, one writes the average potential energy per molecule as

$$\phi_{ave} = -\left(\frac{N}{V}\right) a \quad \text{with} \quad a = 2\pi \int_{r_{min}}^{\infty} \phi(r)r^2 dr \qquad (7.47)$$

Substituting the Lennard-Jones potential and doing the integral yields

$$a = \frac{20\pi r_0^3}{9\sqrt{2}} \phi_{min} \qquad (7.48)$$

So, the van der Waals theory employs two variables a and b and

$$\frac{a}{b} = \frac{5\pi}{9}\phi_{min} \qquad (7.49)$$

The partition function for N particles is approximated as

$$Z_N = K_N \left(\frac{2\pi m}{\beta}\right)^{3N/2} \exp\left(-\beta N \phi_{ave}\right)(V - Nb)^N \qquad (7.50)$$

with $K_N = h^{-3N}/N!$ and $\beta = 1/(kT)$ as before. The free energy is now

$$F = N\phi_{ave} - \frac{N}{\beta}\left[\ln\left(\Lambda\frac{V - Nb}{N}\right) + 1\right] = -\frac{N^2 a}{V} - \frac{N}{\beta}\left[\ln\left(\Lambda\frac{V - Nb}{N}\right) + 1\right] \qquad (7.51)$$

where the 'quantum concentration' Λ is

$$\Lambda = \left(\frac{2\pi m}{\beta h^2}\right)^{3/2} \qquad (7.52)$$

This parameter is related to the 'thermal wavelength' as $\lambda_T = \Lambda^{-1/3}$.

The pressure is found from $p = -\partial F/\partial V$

$$p = -\frac{N^2 a}{V^2} + \frac{N}{\beta(V - Nb)} \qquad (7.53)$$

This is the equation of state for the van der Waals gas. It should be pointed out that while it provides a first-order framework for the understanding of the liquid state in relation to the gaseous state, it is not a very accurate description. This is due to the approximations that have been made and more subtle calculations are available. It must also be pointed out that for realistic fluids the polarity of the molecules will play a substantial role that has been ignored in the van der Waals approach. Nevertheless, analysis of the equation of state gives a broad idea of the physical features that are important. This analysis is done below.

7.7.3 Critical Point of the Van Der Waals Gas

The definition of the *critical point* is the inflection point on the critical isotherm. Keeping the temperature constant this point is found from $\partial p / \partial V = 0$ and $\partial^2 p / \partial V^2 = 0$. Using the equation of state (7.53)

$$2 \frac{N^2 a}{V^3} - \frac{N}{\beta \, (-Nbs + V)^2} = 0; \quad -3 \frac{N^2 a}{V^4} + \frac{N}{\beta \, (-Nb + V)^3} = 0 \qquad (7.54)$$

The critical values that follow from these conditions are

$$V_c = 4Nb; \quad T_c = \frac{8a}{27bk_B}; \quad p_c = \frac{a}{27b^2} \qquad (7.55)$$

Now, expressing the variables in the critical ones (thereby introducing so-called 'reduced' variables) as follows $p = p_c \hat{p}$, $T = T_c \hat{T}$, $V = V_c \hat{V}$, the equation of state is rewritten in the following simple form:

$$\left(\hat{p} \, \hat{V}^2 - 3 \right) \left(3 \, \hat{V} - 1 \right) \hat{\beta} - 8 \, \hat{V}^2 = 0 \qquad (7.56)$$

where $\hat{\beta} = 1/(k_B \hat{T})$. Equation (7.56) represents a cubic equation in $\hat{V}$ and so there is always one solution, but in certain cases, there may be three solutions. Solving for $\hat{p}$ yields

$$\hat{p} = \frac{8 \, \hat{V}^2 - 9 \, \hat{V} \, \hat{\beta} + 3 \, \hat{\beta}}{\hat{\beta} \left(3 \, \hat{V} - 1 \right) \hat{V}^2} \qquad (7.57)$$

This is plotted as the isotherms in Fig. 7.2. On the right-hand side of the figure, the gas state is identified; it is asymptotically characterised by the Boyle-Gay-Lussac law at constant temperature $\hat{p}\hat{V} = cnst$. The left-hand side of the figure, corresponding to higher densities, exhibits entirely different behaviour. Here, the interactive potential plays a crucial role. Note that $V/V_c > 1/3$. $V = V_c/3$ is the volume at which the whole space is taken up with the hard spheres.

The equation of state gives the pressure as a function of prescribed values of the temperature and the volume (in reduced units in the figure). However, prescribing the pressure and the temperature only gives a unique answer for the volume when the temperature is above the critical value T_c. Otherwise, there may be three answers, as illustrated in the figure by the dashed horizontal line. The question is now whether any of these answers is unphysical and should be discarded. It is normally expected that the compressibility of a gas or fluid $\partial \hat{p} / \partial \hat{V}$ is negative. In Fig. 7.2, the intersection

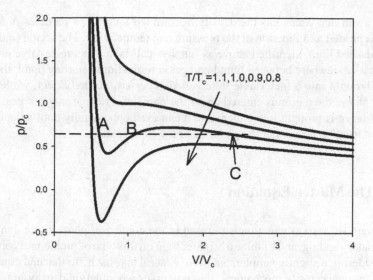

Fig. 7.2 Isotherms for a van der Waals gas

Fig. 7.3 Cusp catastrophy
for a van der Waals gas

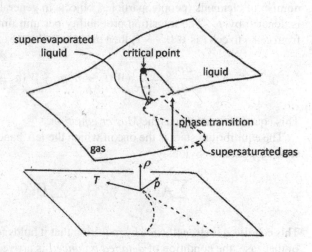

points with an isobar for which there are three solutions for $\hat{V}$ are indicated by A, B and C. The solution marked by point B in the figure does not pertain to a negative compressibility and therefore does not correspond to a physically stable equilibrium state. The two points A and C refer to a high density and low density, respectively, and may be identified with the liquid state and the gas state.

Schematically, further light is shed on this problem. A three-dimensional graph is shown in Fig. 7.3. Such a figure is known as a *cusp catastrophe*; catastrophe theory describes states of systems that have to, in some sense, choose a certain outcome from an unstable intermediate position, Zeeman (1977). There are many applications of this theory, in physics, biology, life sciences, psychology, as well as economics.

For a van der Waals gas the density, for a fixed number of particles N, that is N/V, is plotted as a function of the pressure and temperature. The region (indicated by the dashed line), identified above as 'unphysical' is in fact a metastable physical state; it is somewhere between liquid and gas. From a liquid starting point, the fluid can be brought into a metastable state called *super-evaporated liquid*, while, from the gas phase, the region is entered under the name of a *super-saturated gas*. When the substance is brought into either state it can exist ephemerally until a fluctuation causes it to jump into either the gas state or the liquid state.

7.8   The Master Equation

Boltzmann's transport equation is phrased in the context of phase space dynamics. More understanding, and to shine a brighter light on this type of problems, is obtained by considering a slightly simpler, but more general approach. To that end consider a system, not necessarily mechanical, that possesses a population density in terms of numbers per 'cell'. The cells, or levels, are denoted by a natural number n and the number of elements (people, particles, objects in general) of the system in each cell is denoted by p_n. The transition probability per unit time for one element to jump from cell n to cell i is $W(n \rightarrow i)$ then the rate of change of the population in level n is

$$\frac{dp_n}{dt} = \sum_i \left(W(i \rightarrow n) p_i - W(n \rightarrow i) p_n \right) \tag{7.58}$$

This equation is known as the *Master equation*.

The equilibrium state is the one in which the left-hand side vanishes. Denote this state by p_n^e, then obviously

$$\sum_i \left(W(i \rightarrow n) p_i^e \right) = \sum_i \left(W(n \rightarrow i) p_n^e \right) \tag{7.59}$$

This condition is strengthened by requiring that it holds for each term under the sum. In that case, the condition of *detailed balanced* is arrived at

$$W(i \rightarrow n) p_i^e = W(n \rightarrow i) p_n^e \tag{7.60}$$

Detailed balance can be proven for closed mechanical systems, when Hamiltonian mechanics is valid, when the system is on a micro-scale *reversible*, in other words, if the system is invariant under time reversal. If the principle of detailed balance applies, a Lyapunov function is available. Introduce the function

$$H = \sum_n p_n \log \left(\frac{p_n}{p_n^e} \right) \tag{7.61}$$

Then evaluate its time derivative

$$\frac{dH}{dt} = \sum_n \frac{dp_n}{dt} \left[1 + \log\left(\frac{p_n}{p_n^e}\right) \right] \qquad (7.62)$$

Substitute the Master equation and apply the principle of detailed balance

$$\frac{dH}{dt} = \sum_n \sum_i p_i^e W(i \to n) \left(\frac{p_i}{p_i^e} - \frac{p_n}{p_n^i} \right) \left[1 + \log\left(\frac{p_n}{p_n^e}\right) \right] \qquad (7.63)$$

The summation indices n and i may be swapped and the outcome is the same. Again applying the principle of detailed balance dH/dt is

$$\frac{dH}{dt} = \sum_n \sum_i p_i^e W(i \to n) \left(\frac{p_n}{p_n^e} - \frac{p_i}{p_i^e} \right) \left[1 + \log\left(\frac{p_n}{p_n^e}\right) \right] \qquad (7.64)$$

Adding up gives

$$2\frac{dH}{dt} = \sum_n \sum_i p_i^e W(i \to n) \left(\frac{p_i}{p_i^e} - \frac{p_n}{p_n^e} \right) \log\left(\frac{p_n p_i^e}{p_n^e p_i}\right) \qquad (7.65)$$

This is always negative, unless $p_n = p_n^e$ when it is zero, thus proving the Lyapunov character of the function H.

7.9   The Van Kampen Expansion of the Master Equation

The Dutch physicist van Kampen devised an expansion method for the Master equation. The key idea is that, in many cases, it is possible to distinguish a 'large' parameter. For example, as explained in the beginning of this chapter, for macroscopic systems the number of particles N is a large number. Fluctuations are of the order of $\sqrt{N}$; their effect on the large-scale properties of the system will therefore be of the order of $N^{-1/2}$. Van Kampen (1992), has put forward a general expansion in terms of a large parameter, Ω; the exact form of this parameter is left unspecified for the moment.

It must be pointed out that this is quite an obscure corner of physics, which is very rarely treated in the mainstream literature. However, it will be shown that it is a very important area of study, as it will show clearly how equilibrium of many-particle systems comes into being and also how macroscopic metastability becomes manifest, such as the one demonstrated above for the van der Waals gas.

The parameter Ω is used to distinguish scales. The magnitude of the jumps in the Master equation is denoted by X, so that when Ω is changed the magnitude of X remains the same. In other words, the size of the system is so large that the jumps between states are not affected by it. X refers to microscopic features. The other scale is one that refers to macroscopic changes in the system. These are called x; this change is on a scale of $x = X/\Omega$. x is a concentration.

The Master equation is written in integral form as follows

$$\dot{P}(X,t) = \int \left[W_\Omega(X' \to X) P(X',t) - W_\Omega(X \to X') P(X,t) \right] dX' \qquad (7.66)$$

Now, W is written as a function of the jump length $r = X - X'$: $W_\Omega(X' \to X) = W_\Omega(X - X' \to X') = W_\Omega(r \to X')$. Instead of W another function Φ is introduced that expresses the jump probability in terms of concentrations and the 'jump length' r

$$W_\Omega(X - X' \to X') = \Phi\left(\frac{X'}{\Omega}, X - X'\right) = \Phi(x',r) \qquad (7.67)$$

It follows that $W_\Omega(X \to X') = \Phi(x,-r)$.

The idea is to make an expansion in Ω^{-1}. Note that if the Ws are multiplied by a constant factor then this may be incorporated in the time scale, so, more generally, the form of the Ws, expanded in Ω^{-1} takes the form

$$W_\Omega(X' \to X) = f(\Omega)\left[\Phi_0\left(\frac{X'}{\Omega},r\right) + \Omega^{-1}\Phi_1\left(\frac{X'}{\Omega},r\right) + \Omega^{-2}\Phi_2 + \cdots \right]$$
$$(7.68)$$

This form, called the *canonical form*, is at the heart of the expansion method. When W cannot be cast into this form the method cannot be applied. If, on the other hand, the canonical form *is* applicable it can be substituted in the Master equation to give

$$\frac{\partial P(X,t)}{\partial t} = f(\Omega) \int \left[\Phi_0\left(\frac{X-r}{\Omega},r\right) + \Omega^{-1}\Phi_1\left(\frac{X-r}{\Omega},r\right) + \cdots \right] P(X-r,t)dr$$
$$- f(\Omega) \int \left[\Phi_0\left(\frac{X}{\Omega},-r\right) + \Omega^{-1}\Phi_1\left(\frac{X}{\Omega},-r\right) + \cdots \right] dr P(X,t)$$
$$(7.69)$$

How might a system evolve from an initial, not necessarily equilibrium, state to a state, which, when viewed on a macroscopic scale, is in equilibrium? An illustration in cartoon form, entirely intuitive, is given in Fig. 7.4. Initially, the system is supposed to be in a sharply defined state: $P(X,0) = \delta(X - X_0)$. The state will then evolve due to internal fluctuations and the distribution will become wider. The width will be of the order of $\Omega^{1/2}$. So, set

$$X = \Omega\phi(t) + \Omega^{1/2}\xi \qquad (7.70)$$

The first term is macroscopic and $\phi(t)$ follows the main position of the peak. Below, the idea of Eq. (7.70) will be substituted in the Master equation and it transpires that $P(x,t)$ in the first approximation in ξ does not depend on Ω. This is

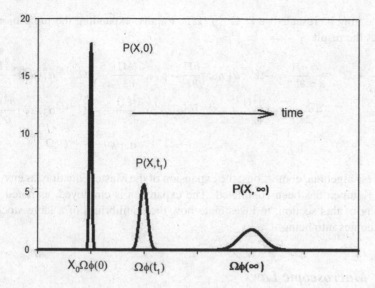

Fig. 7.4 Sketch of the evolution of a system

the crucial element in the van Kampen expansion; it leads to a systematic expansion in $\Omega^{-1/2}$ and *forms the basis for the existence of a macroscopic deterministic description of systems that are inherently stochastic.*

The actual analysis is straightforward. The assumption (7.70) is interpreted as a transformation from X to ξ (a transformation that is dependent on $\phi(t)$). Under the transformation, P is expressed as

$$P(X, t) = P(\Omega\phi(t) + \Omega^{1/2}\xi, t) = \Pi(\xi, t) \tag{7.71}$$

Substituting in the Master equation and expanding and ordering the terms by the powers of $\Omega^{-1/2}$ leads to

$$\frac{\partial \Pi(\xi, t)}{\partial t} - \Omega^{1/2}\frac{d\phi}{dt}\frac{\partial \Pi}{\partial \xi} = -f(\Omega)\int\left[\Phi_0\left(\frac{X}{\Omega}, -r\right) + \Omega^{-1}\Phi_1\left(\frac{X}{\Omega}, -r\right) + \cdots\right]dr\, P(X, t)$$

$$= -\Omega^{-1/2}f(\Omega)\frac{\partial}{\partial \xi}\int r\Phi_0(\phi(t) + \Omega^{-1/2}\xi, r)dr.\Pi(\xi, t) +$$

$$+\frac{1}{2}\Omega^{-1}f(\Omega)\frac{\partial^2}{\partial \xi^2}\int r^2\Phi_0(\phi(t) + \Omega^{-1/2}\xi, r)dr.\Pi(\xi, t) +$$

$$-\frac{1}{3!}\Omega^{-3/2}f(\Omega)\frac{\partial^3}{\partial \xi^3}\int r^3\Phi_0(\phi(t) + \Omega^{-1/2}\xi, r)dr.\Pi(\xi, t) +$$

$$-\Omega^{-3/2}f(\Omega)\frac{\partial}{\partial \xi}\int r\Phi_1(\phi(t) + \Omega^{-1/2}\xi, r)dr.\Pi(\xi, t) + O(\Omega^{-2}) \tag{7.72}$$

The notation is made more compact by introducing the 'jump moments'

$$\alpha_{\nu,\lambda}(x) = \int r^\nu\Phi_\lambda(x, r)dr \tag{7.73}$$

The time may be rescaled: $\tau = \Omega^{-1} f(\Omega) t$. Finally, expanding the jump moments produces the result

$$\frac{\partial \Pi(\xi, \tau)}{\partial \tau} - \Omega^{1/2} \frac{d\phi}{d\tau} \frac{\partial \Pi}{\partial \xi} = -\Omega^{1/2} \alpha_{1,0}(\phi) \frac{\partial \Pi}{\partial \xi} - \alpha'_{1,0} \frac{\partial (\xi \Pi)}{\partial \xi} - \frac{1}{2} \Omega^{-1/2} \alpha''_{1,0}(\phi) \frac{\xi^2 \Pi}{\partial \xi} +$$

$$+ \frac{1}{2} \alpha_{2,0}(\phi) \frac{\partial^2 \Pi}{\partial \xi^2} + \frac{1}{2} \Omega^{-1/2} \alpha'_{2,0}(\phi) \frac{\partial^2 \xi \Pi}{\partial \xi^2} - \frac{1}{3!} \Omega^{-1/2} \alpha_{3,0}(\phi) \frac{\partial^3 \Pi}{\xi^3} +$$

$$- \Omega^{-1/2} \alpha_{1,1}(\phi) \frac{\partial \Pi}{\partial \xi} + O(\Omega^{-1}) \qquad (7.74)$$

After these algebraic contortions, the expansion of the Master equation, as envisaged by van Kampen has been completed. The expansion is employed, as noted at the beginning of this section, to investigate how the equilibrium of a large stochastic system comes into being.

7.9.1   Macroscopic Laws

When inspecting the expansion, Eq. (7.74) an at-first-sight disturbing feature is observed. It is that an expansion in $\Omega^{-1/2}$, for large Ω, should include terms that are proportional to $\Omega^{1/2}$, which blow up. It is seen that both these terms are proportional to Π and therefore on requiring that they cancel the infinity is avoided. The implication is that

$$\frac{d\phi}{d\tau} = \alpha_{1,0}(\phi) \qquad (7.75)$$

Now, ϕ is the function that determines the evolution of the system, as illustrated in Fig. 7.4, so Eq. (7.75), through the functional dependence of $\alpha_{1,0}(\phi)$, will describe whether the system is stable or not.

Various scenarios for stability can be sketched, see Fig. 7.5. In the cartoon labelled a, a stable system is indicated. The time dependence, through Eq. (7.75), always ends up in $\phi(\infty) = \phi_s$. Similarly, the hodogram in b represents an unstable system, in spite of the fact that $d\phi/d\tau = 0$ when $\alpha_{1,0}(\phi_s) = 0$. An alternative scenario that does not lead to stability is sketched in Fig. 7.5c. The situation that qualitatively looks like Fig. 7.5d has two stable and one unstable point. These are all *macroscopic* features.

The stable case, as sketched in Fig. 7.5a can be explored further. It is clear that there is always stability when $\alpha'_{1,0}(\phi) \le 0$ for all ϕ. Somewhat stronger, $\alpha'_{1,0}(\phi) \le -h < 0$, it assures that stability is reached exponentially always faster than $e^{-h\tau}$. The stability is global.

It becomes particularly interesting when $\alpha_{1,0}$ can be derived from a potential: $\alpha_{1,0} = -\partial V(\phi)/\partial \phi$. Now the stability can be investigated by considering the time dependence of the potential. Using (7.75)

$$\frac{dV}{dt} = \frac{dV}{d\phi} \frac{d\phi}{dt} = -\alpha^2_{1,0} \le 0 \qquad (7.76)$$

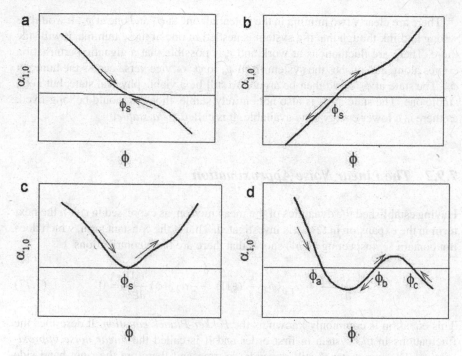

Fig. 7.5 Various possible scenarios for stable and unstable systems

Furthermore, as it follows from the condition $\alpha'_{1,0} < 0$ that $V''(\phi) > 0$, it is seen that the potential $V(\phi)$ acts as a function of Lyapunov. The equal sign holds when $\alpha_{1,0}(\phi) = 0$, which is in $\phi = \phi_s$. Not all stability follows from this, of course. For example, the point $\phi = \phi_c$ in Fig. 7.5d is stable, but in that case, there is no function of Lyapunov to underpin it.

Even if there is no function of Lyapunov, the concept of a potential still appeals. Taking the case of Fig. 7.5d, a potential can be obtained. This potential would look like the sketch in Fig. 7.6.

Fig. 7.6 Sketch of the potential for the case of Fig. 7.5d

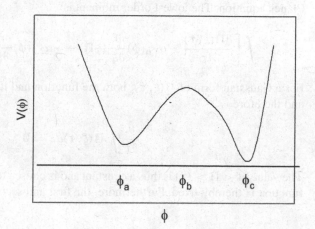

There are clearly two minima in the potential, one at ϕ_a and one at ϕ_c. It would be wrong to think though that if a system is nestled in one of these minima, it will stay there. There are fluctuations at work and it is possible that a gigantic perturbation comes along and pushes the system from ϕ_a to ϕ_c or vice versa across the hump at ϕ_b. The state at ϕ_b could then be argued to still be a viable physical state, but not a stable one. The state at ϕ_a is also not entirely stable, though it could be long-lived, as there is a lower energy state available. It is called *meta-stable*.

7.9.2   The Linear Noise Approximation

Having established the dynamics of the mean motion, as expressed in $\phi(\tau)$, the next term in the expansion in $\Omega^{-1/2}$ is investigated. That is the constant term, which does not contain Ω. Inspecting (7.74) shows that there are three contributions

$$\frac{\partial \Pi(\xi, \tau)}{\partial \tau} + \alpha'_{1,0}(\phi)\frac{\partial}{\partial \xi}(\xi \Pi) - \frac{1}{2}\alpha_{2,0}(\phi)\frac{\partial^2 \Pi}{\partial \xi^2} = 0 \qquad (7.77)$$

This equation is commonly known as the *Fokker-Planck equation*. It describes the fluctuations in the system in first order and it is called the *linear noise approximation*. The parameter ϕ still depends on time and therefore the right-hand side has coefficients that are time-dependent. A form of a solution is readily found; put forward

$$\Pi(\xi, \tau) = \exp\left(D(\tau) + E(\tau)\xi + F(\tau)\xi^2\right) \qquad (7.78)$$

where $D(\tau)$, $E(\tau)$ and $F(\tau)$ are yet to be determined functions. Substituting this form in the Fokker-Planck equation yields a quadratic polynomial in ξ, as is easily verified. The three terms give rise to first-order coupled differential equations in the three functions. It follows then that the solution has a Gaussian form. The heavy work of determining the precise form of the three differential equations is not necessary, as the moments of the distribution $\Pi(\xi, \tau)$ can be read immediately from the Fokker-Planck equation. The lowest order moment is

$$\int \left[\frac{\partial \Pi(\xi, \tau)}{\partial \tau} + \alpha'_{1,0}(\phi)\frac{\partial}{\partial \xi}(\xi \Pi) - \frac{1}{2}\alpha_{2,0}(\phi)\frac{\partial^2 \Pi}{\partial \xi^2}\right] d\xi = 0 \qquad (7.79)$$

For a Gaussian form of $\Pi(\xi, \tau)$, both the function and its derivative vanish at $\pm\infty$ and therefore

$$\frac{\partial}{\partial \tau}\int \Pi(\xi, \tau)d\xi = 0 \qquad (7.80)$$

The value of $< \Pi > (\tau)$ is thus a constant and is chosen to be equal to unity; the first function is thereby fixed. Furthermore, the first and second moments refer directly to the average and standard deviation

$$< \xi > (\tau) = \int \xi \Pi(\xi, \tau) d\xi; \quad < \xi^2 > (\tau) = \int \xi^2 \Pi(\xi, \tau) d\xi \qquad (7.81)$$

The time derivatives of these quantities are easily read from the Fokker-Planck equation

$$\frac{\partial < \xi >}{\partial \tau} = \alpha'_{1,0}(\phi) < \xi >; \quad \frac{\partial < \xi^2 >}{\partial \tau} = 2\alpha'_{1,0}(\phi) < \xi^2 > + \alpha_{2,0}(\phi) \qquad (7.82)$$

It was seen that a stable equilibrium can be reached when $\alpha'_1 < 0$. At equilibrium, the rate of change of the macroscopic variables $< \xi >$ and $< \xi^2 >$ vanishes so that, in that case,

$$\frac{\partial < \xi >}{\partial \tau} = 0 \rightarrow < \xi >_s = 0; \quad \frac{\partial < \xi^2 >}{\partial \tau} = 0 \rightarrow < \xi^2 >_s = -\frac{\alpha_{2,0}(\phi_s)}{2\alpha'_{1,0}(\phi_s)} \qquad (7.83)$$

In interpreting these findings physically, it is recalled that ξ denotes how far a microscopic state variable X has strayed from its average value (scaled by a factor $\Omega^{1/2}$) and the jump moment α informs, loosely speaking, on an average jump probability weighed with the jump distance. So, (7.83) makes sense, as α_2 says something about the scatter in the jump probabilities (if it is small the jumps are more likely to be to nearby states and if it is large the jumps to faraway states are also quite probable). α'_1 was seen to relate to the kind of process that the average approaches to system stability. If the slope to a stable equilibrium in Fig. 7.5a is steep then the system gets to the stable point very quickly and Eq. (7.83) expresses the fact that the mean scatter in the states is smaller than when the slope is less steep. Intuitively, these results make sense. Van Kampen's (1921–2013) achievement was that he managed to express it in mathematically rigorous terms.

7.9.3   Diffusion

The treatment so far has been analogous to the concepts that underlie statistical mechanics, namely that there is a thermal equilibrium that corresponds to the most probable distribution of energy states. In the preceding section, it was seen how this comes about: $\alpha'_1 < 0$. What happens in the limiting case when $\alpha'_1 = 0$? Naively, it would follow from Eq. (7.82) that $< \xi^2 > = \alpha_{2,0}\tau$. That would mean that the scatter in the states would always expand and the cartoon view as expressed in Fig. 7.4, where the state at $\tau \rightarrow \infty$ is in equilibrium with a limited fluctuation content, is no longer valid. That makes the assumptions about the expansion of the Master equation invalid as well. Nevertheless, in physics, there are processes that have the property of always expanding fluctuations. An alternative expansion has to be found to accommodate that situation.

For these processes, it may be assumed that, after they have progressed for a while, the probability P will be a smoothly varying function of the internal states X with a width of the large parameter Ω. The natural expansion parameter is therefore the concentration $x = X/\Omega$ and the Master equation may be written in these terms, completely analogously to the way it was done in the previous sections. Following van Kampen (1992), again

$$\frac{\partial P(X,t)}{\partial t} = f(\Omega) \int \left[\Phi_0\left(x - \frac{r}{\Omega}, r\right) + \Omega^{-1}\Phi_1\left(x - \frac{r}{\Omega}, r\right) + \cdots \right] P\left(x - \frac{r}{\Omega}, t\right) dr$$
$$- f(\Omega) \int \left[\Phi_0(x, -r) + \Omega^{-1}\Phi_1(x, -r) + \cdots \right] dr\, P(x,t)$$
(7.84)

Using the definition of α as before now results in

$$\frac{\partial P(x,t)}{\partial t} = \Omega^{-2} f(\Omega)\left[-\frac{\partial(\alpha_{1,1}(x)P)}{\partial x} + \frac{1}{2}\frac{\partial^2(\alpha_{2,0}(x)P)}{\partial x^2} \right] +$$
$$+\Omega^{-3} f(\Omega)\left[\frac{1}{2}\frac{\partial^2(\alpha_{2,1}(x)P)}{\partial x^2} - \frac{1}{3!}\frac{\partial^3(\alpha_{3,0}(x)P)}{\partial x^3} - \frac{\partial(\alpha_{2,1}P)}{\partial x} \right] + O(\Omega^{-4})$$
(7.85)

Introducing a new timescale $\tau = t\Omega^{-2} f(\Omega)$, the lowest order is the top line in Eq. (7.85). It is the non-linear Fokker-Planck equation and is known as the *diffusion approximation*

$$\frac{\partial P(x,\tau)}{\partial \tau} = -\frac{\partial(\alpha_{1,1}(x)P)}{\partial x} + \frac{1}{2}\frac{\partial^2(\alpha_{2,0}(x)P)}{\partial x^2}$$
(7.86)

There is now no 'motion of the average', that is, an equation for $\phi(t)$. The only information is the fluctuation content as it is being distributed over the space.

7.9.3.1   Constant Diffusion Coefficient

A special case of the diffusion approximation is one that results in the so-called *heat equation*. For such processes, $\alpha_{1,1} = 0$ and $\alpha_{2,0}$ is a constant. This equation appears in a large number of *transport processes*: heat transport, diffusion, viscous processes including consolidation phenomena (in soil mechanics). The equation is mercifully linear. The coefficient $\frac{1}{2}\alpha_{2,0}$ will be called D.

The heat equation in one dimension reads ($u(x,t)$ is the temperature and D the *(thermal) diffusivity* or the *thermometric conductivity*)

$$\frac{\partial u}{\partial t} = D\frac{\partial^2 u}{\partial x^2}$$
(7.87)

For three-dimensional applications the second derivative is replaced by ∇^2; for illustration purposes in order to understand the character of the solutions, the one-dimensional version will be studied. A *similarity solution* is put forward. Let $u(x, t) = At^a v(t^b x^c)$. Substituting in the heat equation gives

$$- Dt^{2b+1} v'' + ybv' + av = 0 \tag{7.88}$$

which has a solution of the form $v = e^{-dy^2}$ and it follows that

$$a = -\frac{1}{2}; \quad d = \frac{1}{4D}; \quad b = -\frac{1}{2}; \quad u(x, t) = A\frac{1}{\sqrt{t}}\exp\left(-\frac{x^2}{4Dt}\right) \tag{7.89}$$

Now, it is noticed that the spatial differentiation is of the second order and therefore a translation over an amount x' leaves the equation invariant, so the following are also solutions:

$$u(x, t) = A\frac{1}{\sqrt{t}}\exp\left(-\frac{(x - x')^2}{4Dt}\right); \quad u(x, t) = \frac{A}{\sqrt{t}}\int_{-\infty}^{\infty} f(x')\exp\left(-\frac{(x - x')^2}{4Dt}\right) dx' \tag{7.90}$$

where $f(x')$ is a 'decent' function and can be related to the value of $u(x, 0)$ as follows. Change the integration variable to $x' = x + 2y\sqrt{Dt}$, then

$$u(x, t) = A\int_{-\infty}^{\infty} f\left(x + 2y\sqrt{Dt}\right)\exp\left(-y^2\right) dy \tag{7.91}$$

The limit $t \to 0$ can now be taken

$$u(x, 0) = A\int_{-\infty}^{\infty} f(x)\exp\left(-y^2\right) dy = A\sqrt{\pi} f(x) \tag{7.92}$$

This establishes the connection between the initial value of u and the function $f(x)$.

As an example consider a source at $t = 0$ of size $2a$ where the temperature (the value of u) is u_0. So, $f(x) = u_0/\sqrt{\pi}$ if $-a \leq x \leq a$. Doing the integral gives for the temperature as a function of position and time

$$u(x, t) = \frac{1}{2}\left[\text{erf}\left(\frac{a - x}{2\sqrt{Dt}}\right) + \text{erf}\left(\frac{a + x}{2\sqrt{Dt}}\right)\right] \tag{7.93}$$

The process is illustrated in Fig. 7.7. The region $x > 0$ is shown only, as the diffusion is symmetrical in x.

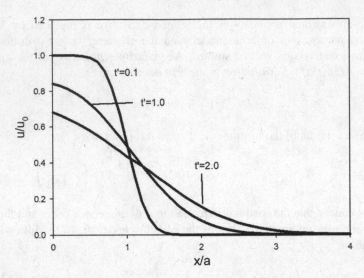

Fig. 7.7 Diffusion from a source at $-a < x < a$ at various times: $t' = 4tD/a^2$

7.9.3.2 Heat Flow in the Earth's Crust

1. The heat equation can be solved for a half-space $x > 0$ under the condition that $u(x, 0) = u_0$ for $x > 0$, while the temperature at $x = 0$ is being kept constant at $u(0, t) = 0$. The idea is that when the Earth was formed it was a liquid ball that subsequently cooled and solidified. Solving the heat equation then gives the temperature gradient at the surface of the sphere, which can be measured and from that—and known heat conduction properties—the *age of the Earth* can be estimated. These considerations were put forward by Fourier and Kelvin.

The easiest way of getting a solution for this problem is by Laplace transform. The transformed equation reads

$$s\hat{u}(x) - u_0\theta(x) = D\frac{\partial^2 \hat{u}(x)}{\partial x^2} \qquad (7.94)$$

Here, $\theta(x) = 1$ for $x > 0$ and $\theta(x) = 0$ for $x < 0$. The solution for $x > 0$, enforcing the condition $u(0, t) = 0$ yields

$$\hat{u}(x) = \frac{u_0}{s} - \frac{u_0}{s}\exp\left(-\sqrt{\frac{s}{D}}x\right) \;\;\rightarrow\;\; u(x, t) = u_0\mathrm{erf}\left(\frac{x}{2\sqrt{Dt}}\right) \qquad (7.95)$$

Kelvin in 1862 (Thomson 1862), used this to estimate the age of the Earth, see the discussion in Carslaw and Jaeger (1959), where a large number of analytical solutions for the heat equation are presented. The geothermal gradient can be measured (though it varies widely from place to place) and Kelvin took the temperature u_0 to be the one of melting rock: 4000 °C. For D, he took 0.0118 cm^2/s. He arrived at the estimated age of the Earth, that is the point at which solidification began, as of the order of 10^8 years. This is a good factor of ten out from the more modern estimates, which put forward the number of 4.45×10^9 years. Kelvin's calculation can be much refined, but is essentially wrong, because the underlying assumption is that there are no heat sources—radioactivity, chemical reactions, etc.—accounted for. Of course, at the time radioactivity was not yet known.

2. The *penetration of heat into the soil* is analysed by solving the heat equation for a half-space $x > 0$ under the boundary condition $u(0, t) = \bar{u} + A(0)\cos\omega t$. The solution has the form $u(x, t) = \bar{u} + A(x)\exp(i\omega t)$. Substituting in the heat equation shows that the solution is

$$u(x, t) = \bar{u} + A(0)\exp\left(-\sqrt{\frac{\omega}{2D}}x\right)\cos\left(\sqrt{\frac{\omega}{2D}}x - \omega t\right) \qquad (7.96)$$

The *penetration depth* is $\pi\sqrt{2D/\omega}$, which is the depth at which the amplitude is 95% of the surface value. The penetration depth depends on the frequency of the surface fluctuation, so for diurnal variations the penetration is much less pronounced than for seasonal variations. It also depends on the type of soil. For example, there is a pronounced difference between the behaviour of a sandy soil and a peat; the latter conducts better as it usually has a higher moisture content.

For a dry, sandy soil $D = 0.0020\,\mathrm{cm}^2/\mathrm{s}$; for a moist, sandy soil $D = 0.0045\,\mathrm{cm}^2/\mathrm{s}$. It follows that for diurnal variations in temperature the penetration depth in dry, sandy soil is some 23 cm, while, for an annual variation, it is approximately 4.5 m.

In environmental science, this subject is important, because the extent of the temperature penetration determines how much locked-in gas will be released, which is relevant for the assessment of the atmospheric greenhouse effect.

References

Carslaw HS, Jaeger JC (1959) Conduction of heat in solids, 2nd edn. Clarendon Press, Oxford

Fermi E (1956) Thermodynamics. Dover, New York

Jui Sheng H (1975) Principles of thermodynamics. McGraw-Hill Kogakusha, Tokyo

Landau LD, Lifshitz EM (1976) Mechanics (Course on theoretical physics, Volume 5). Pergamon, Oxford

Mandl F (1988) Statistical physics, 2nd edn. Wiley, Chichester

Pointon AJ (1967) Introduction to statistical physics. Longman, London

Pendlebury JM (1985) Kinetic theory. Adam Hilger, Bristol

Saggion A, Faraldo R, Pierno M (2019) Thermodynamics: fundamental principles and applications. Springer, Cham

Ter Haar D (1966) Elements of thermostatistics, 2nd edn. Holt, Rinehart and Winston, New York

Thomson W (1862) On the secular cooling of the earth. Earth Environ Sci, Trans R Soc Edinb 23(1):157–169. https://doi.org/10.1017/S0080456800018512

Van Kampen NG (1992) Stochastic processes in physics and chemistry. North Holland (Elsevier), Amsterdam

Zeeman EC (1977) Catastrophe theory. Addison-Wesley, London

Chapter 8
Elements of Quantum Mechanics

Abstract Some introductory and historical remarks arguing the necessity of quantum mechanics are followed by a description of material waves describing free particles. Heisenberg's uncertainty relation is demonstrated and Schrödinger's equation is made plausible. Examples of the solution of Schrödinger's equation are shown: a particle in a well, the hydrogen atom and the harmonic oscillator. Links to classical mechanics are shown. Dirac's ideas are introduced and an operator algebra on ket vectors is put forward. A long section on quantum measurement is presented in order to understand better how different quantum mechanics is from classical mechanics. This culminates in a discussion of the violation of the Bell inequalities.

8.1 Introductory Remarks

The subject of quantum mechanics is less than a century and a half old. Yet, it is one of the most extensive branches of physics, incorporating molecular, solid state, atomic, nuclear and elementary particle physics. As such it describes all of spectroscopy, many branches of optics, quite a lot of chemistry, electronics, superconductivity, aspects of astronomy (including stellar physics) and certain biological applications. This list makes clear that it is an enormous field of intellectual exploration and discussions on all levels, ranging from the very (philosophical) foundations of the subject to the high-dimensional mathematical constructs of string theory. A consequence of the breadth of applications of the subject has meant that there are quite a number of communities that use 'quantum language', each for their own ends. Thus, diverging 'dialects' have developed; it is improbable, for example, that the chemistry community will find that they have much in common with the string theory or elementary particle community and the philosophers definitely speak a different quantum dialect than the electronics engineers. Yet they all 'speak quantum' and avail themselves of particular concepts and have developed an intuition for the logic of the subject, which is both different from and overlaps with classical mechanics. In this section, particular attention will be focussed on the quantum equivalent of the classical problems that have been discussed in the previous chapters, so the differences and correspondences between the two can be elucidated.

M. A. C. Koenders, *Constructing the Edifice of Mechanics*, Undergraduate Texts
in Physics, https://doi.org/10.1007/978-3-031-34071-0_8

From the outset, it must be stated that quantum mechanics has been a very successful theory, not only in the breadth of the topics that it deals with but also in terms of the accuracy of its predictions. There are many excellent introductory books on quantum mechanics (Bes 2013; Gasiorowicz 1995; Goswami 1992; Landau 1976; McMurry 1994; Merzbacher 1998; Sakurai 1985, some are classical texts) and these deal to a large extent with the calculation methods that are required to obtain quantitative answers to specific problems. There is no point in trying to cover that field again, or even trying to improve on it. The approach here is classical in the sense that relativistic quantum mechanics—as developed by Dirac—though very interesting and elegant, is not considered. (The associated field theory, which is extremely powerful and relevant to high-energy physics, is also not studied.) By keeping to the non-relativistic theory, it is possible to look into the foundations of the subject and make a connection with the intuitive ideas of classical mechanics.

Why is quantum mechanics necessary? Following Planck's idea that light can have a corpuscular character, de Broglie came up with the concept that subatomic particles can have wave-like properties. A number of experiments would appear to confirm that and, indeed, Planck's hypothesis was immediately confirmed by experiment through the radiation law. Other early applications, such as Compton scattering and the photoelectric effect, confirmed the basic wave principle.

The most telling break-down of classical mechanics on the subatomic scale is the classical description of an atom. Electrons that circle the nucleus in planetary orbits *must* radiate, as they are charged particles. In so doing, they lose energy and in time every electron would be in such a low orbit that it would collide with the protons. Classical mechanics therefore cannot describe a stable atom. The framework of quantum mechanics *does* provide for stable atoms. However, the classical concept of planetary orbits of electrons around the nucleus has to be abandoned. In its place, an alternative theory that still employs the classical concepts, such as energy and momentum, is constructed. In quantum physics, however, atoms can still radiate light, but the radiation is highly regulated and confined to certain jumps between *discrete energy levels* of the electrons as they are in stationary orbits around the nucleus. The fact that only a certain spectrum of energy levels is possible has lent its name to the subject: the energies are said to be *quantised*. Quantum mechanics successfully describes which levels exactly are possible and, in so doing, it can make accurate predictions of the properties of the light that is radiated when the electron(s) jump from one orbit to another.

In the 1930s, it was discovered that not just energy levels appear to be quantised; particles themselves may have *intrinsic* properties that may be quantised as well. (An example is the property of particle spin, a property that has far-reaching consequences for atomic physics and also—strangely—for statistical mechanics.) This discovery (as well as subsequent ones that have esoteric names such as 'strangeness', 'charm', 'top' and 'bottom') and the tremendous expansion of particle physics, especially after the second world war, has led to an exciting adventure of experiments and theoretical development, which is still continuing to this day. These intrinsic properties define vast zoos of 'elementary' particles that some say form the basic ingredients of the Universe. It is a journey of discovery that is by no means finished yet and which takes place in large particle accelerators and in ever more ingenious astronomical observatories. In this chapter, the intrinsic quantum properties of particles will not be treated; this is not a text on particle physics. The focus is on *mechanics* and especially—as stated above—the link between quantum mechanics and classical mechanics.

The theory begins with the description of matter waves for free particles. Then extensions to particles in a potential field are introduced. The resulting theory describes atomic physics very well; nevertheless, despite its success, the interpretation of the theory as it relates to experiment is for many still obscure. In the following sections, this issue will be studied more closely.

Quantum theory is built up, starting with free particles. If a free-particle theory is to be tested, it needs an experiment. During any experiment, the particles need to interact with a measurement apparatus, implying that, at the moment of measurement, they are not free. The best that can be done in an experiment is to say something about where particles have been. Studying what they are doing while they are free is impossible. Problems like these (and more sophisticated versions of it) explain why the measurement of 'quantum processes' has been—and still is—the subject of intense debate. Conceptually, the problem is that there is a very strong influence from a highly successful long-standing theory that has been tested to an exceptionally high degree of satisfaction: *classical mechanical theory*. Now, obviously, if one lets go of the concept of a corpuscular particle then some of the classical world view has to be abandoned and replaced with something else (this was very hard to accept for some physicists in the first half of the twentieth century, see McCormmach (1982) for a fictional—but not unrealistic—account). The question is: how much can acceptably be retained, or—alternatively—what needs to be thrown out? For example, should the inertial principle be retained, which would imply that concepts like momentum are still valid. Should the concept of a force be retained, so that Newton's second law can continue to be used, or does the wave description (even for a free particle that feels no forces) necessarily mean that a new equation of motion is required? If that is so, then it should be made clear how a *classical limit* follows. Also, Newtonian mechanics gives definite answers to problems: positions and velocities are well-determined properties, open to experimental verification. Is that too much to ask from a theory? Should physical theories be permitted to be vaguer?

Here are a number of problems that come into play when the theory of mechanics is re-examined to allow for particles to be described as waves. While it is one thing to want to describe light as particles, because light has a special place in physics anyway, it is quite another to give wave-like properties to particles. These, at first sight, would seem to be perfectly normal objects, open to the theory of classical mechanics. However, in physics experiment is absolute monarch and experiment leads unambiguously to a wave-like description of subatomic particles. One such experiment, for example, is crystal diffraction, which seems to work as well with light as with electrons and neutrons and for which wave-like properties are essential to come to a satisfactory description.

8.2 Basic Free Particle Wave Properties

Describing particles propagating as waves requires the introduction of a wavelength and a frequency. Free particles have some classical properties that can be adapted to allow for these. Planck's idea of endowing light, or, more generally, electromagnetic radiation, with corpuscular properties to create individual 'light particles' or *photons* involves the introduction of a natural constant h. This constant is *not* part of classical mechanics and in the realm of classical mechanics it is very small: $h = 6.62607015 \times 10^{-34}$ Js. Planck suggested that the energy of a photon is related to its frequency (colour) as $E = hf$. From special relativity theory it follows immediately that the momentum of a photon is related to the wavelength λ as $p = h/\lambda$. Now, free particles have the properties of energy and momentum and de Broglie's idea was to describe them as such. Instead of h the quantity $\hbar = h/(2\pi)$ is used; this is called the *reduced Planck constant*. The wave vector $\mathbf{k}$ is related to the momentum of the particle as $\mathbf{p} = \hbar\mathbf{k}$. Analogously to the photon the energy is related to the circular frequency as $E = \hbar\omega$. In passing it is noted that the wave velocity is $\mathbf{v} = \partial\omega/\partial\mathbf{k}$. It follows that the wave velocity for free particles is $\mathbf{p}/m$, which is the particle velocity. However, in quantum mechanics the velocity is not entirely the same as it is in classical physics. This issue will be elucidated below.

8.2.1 The Necessity of Complex Numbers in Quantum Mechanics

Free particles move in a straight line, so in one dimension the wave would be described as

$$\psi(x, t) = \cos(kx - \omega t) + \delta \sin(kx - \omega t) \tag{8.1}$$

The amplitude, which is another wave property, is currently not yet specified; that will be done below.

An arbitrary displacement in the x-direction or an arbitrary time shift should not change the physical character of the wave. Phase constants should also have no influence on the outcome and so

$$\cos(kx - \omega t + \phi) + \delta \sin(kx - \omega t + \phi) = \tilde{A}(\phi)\left[\cos(kx - \omega t) + \delta \sin(kx - \omega t)\right] \tag{8.2}$$

This is valid for all x and t, hence

$$\cos\phi + \delta\sin\phi = \tilde{A}; \quad \delta\cos\phi - \sin\phi = \tilde{A}\delta \tag{8.3}$$

It follows that $\delta^2 = -1$ or $\delta = \pm i$. Choosing the plus sign leads to $\tilde{A} = e^{i\phi}$. The important conclusion must be that describing particles as waves implies that the wave function is, generally speaking, a fundamentally *complex* quantity, which is a radical departure from classical physics. The wave function for a particle moving in the positive x direction will now have the form

$$\psi(x,t) = Ae^{i(kx-\omega t)} \tag{8.4}$$

8.3 Wave Packets and Uncertainty Relation

One of the wonderful things that can be done with waves is make superpositions. A mix of a number of waves can be put together. This is a first step towards creating a more classical view in which a particle has a *localised identity*. A 'packet' of waves is a superposition of wave functions with different wave vectors $\mathbf{k}$. Let this mix be defined by a weighting function of momenta at time $t = 0$. Given the form of (8.4) it is seen that the momentum weighting function is the Fourier transformed of the spatial wave function. So, momentum and spatial representations contain the same information. Going to three dimensions, call the momentum weighting function $\phi(\mathbf{k})$, then

$$\psi(\mathbf{r},0) = \frac{1}{(2\pi)^{3/2}} \int d^3k\, \phi(\mathbf{k}) e^{i\mathbf{k}\cdot\mathbf{r}}; \quad \phi(\mathbf{k},0) = \frac{1}{(2\pi)^{3/2}} \int d^3r\, \psi(\mathbf{r}) e^{-i\mathbf{k}\cdot\mathbf{r}} \tag{8.5}$$

Where no integral boundaries are indicated the range is $(-\infty,\infty)$.

Now, for simplicity, go back to one dimension and assume that the momentum weighting function is of a form that is centred on $\bar{p}$ with a width Δp and has a Gaussian shape

$$\phi(p) = Ce^{-(p-\bar{p})^2/\Delta p^2} \tag{8.6}$$

The position-dependent wave function is evaluated by doing the integral in the first of Eq. (8.5) and has the form

$$\psi(x,0) = \frac{C\Delta p^2}{\sqrt{2}\hbar} e^{\frac{i}{\hbar}\bar{p}x} \exp\left(-\frac{\Delta p^2 x^2}{4\hbar^2}\right) \tag{8.7}$$

This represents a wave that propagates with momentum $\bar{p}$ and has a Gaussian shape with width $\Delta x = 2\hbar/\Delta p$. This is no longer a wave that propagates from $-\infty$ to $+\infty$, but a *localised phenomenon* within a region of the order of Δx. The widths of the two distributions are related as

$$\Delta p \Delta x \approx 2\hbar = \frac{h}{\pi} \tag{8.8}$$

The precise form of the distribution is not so important: any limited momentum distribution will lead to a localised wave function and the product of the uncertainties in momentum and position is always in the order of magnitude of Planck's constant. This result is known as *Heisenberg's uncertainty relation*.

It should now be clear that describing particles as wave-like phenomena leads to outcomes that stray far from classical mechanics. While concepts from Newtonian physics, such as position and momentum, have not been abandoned, the notion of uncertainty is definitely not a classical one.

Further probing into the theory becomes even more unsettling for the classical physicist. The question may be asked: if there is an uncertainty in position can anything be said about where the particle actually is? The answer to this question can only be provided using probabilistic concepts. In the above example the most likely position is that the particle is at $x = 0$, however other positions are not impossible. In principle the particle could be at $\pm\infty$, though that is very unlikely. Remember that the wave function is fundamentally complex (C could be a complex number), a quantitative answer to the question 'where is the particle?' must be a real quantity. It is therefore not unreasonable to identify the *intensity* of the wave function with a *probability*. In other words $\psi(x)\psi^*(x)dx$ may be identified with the probability to find the particle in a region dx at the location x. Up until now the amplitude of the wave has been left vague. However, realising that the particle must be somewhere between $-\infty$ and $+\infty$ implies that the amplitude of the wave function must satisfy the relation

$$\int \psi(x)\psi^*(x)dx = 1 \tag{8.9}$$

The average position is (with apologies for the strange ordering of the terms)

$$< x > = \frac{\int \psi^*(x)x\psi(x)dx}{\int \psi^*(x)\psi(x)dx} = \int \psi^*(x)x\psi(x)dx \tag{8.10}$$

The same reasoning can be employed for the momenta. The momentum representation is completely equivalent to the coordinate representation, so

$$< p > = \frac{\int \phi^*(p)p\phi(p)dp}{\int \phi^*(p)\phi(p)dp} = \int \phi^*(p)p\phi(p)dp \tag{8.11}$$

To recast this into the coordinate representation use the fact that

$$\phi(p) = \frac{1}{\sqrt{\pi}} \int dx\psi(x)e^{-\frac{i}{\hbar}px} \tag{8.12}$$

Then

$$p\phi^*(p) = \frac{1}{\sqrt{\pi}} \int dx\psi^*(x)\left(\frac{\hbar}{i}\right)\frac{\partial}{\partial x}e^{\frac{i}{\hbar}px} \tag{8.13}$$

And

$$< p > = \int dp\phi(p)p\phi^*(p) = \int dx\psi^*(x)\left(\frac{\hbar}{i}\right)\frac{\partial}{\partial x}\psi(x) \tag{8.14}$$

Entirely analogously the analysis can be done in three dimensions to give

$$< \mathbf{p} >= \int d^3 p \, \phi(\mathbf{p}) \mathbf{p} \phi^*(\mathbf{p}) = \int d^3 x \, \psi^*(\mathbf{x}) \left(\frac{\hbar}{i} \right) \nabla \psi(\mathbf{x}) \qquad (8.15)$$

The complex conjugate of the integrand is $-\psi(\mathbf{x}) \, (\hbar/i) \, \nabla \psi^*(\mathbf{x})$. As $< \mathbf{p} >$ is real, the velocity density is

$$\mathbf{j} = \frac{1}{2m} \left[\psi^*(\mathbf{x}) \left(\frac{\hbar}{i} \right) \nabla \psi(\mathbf{x}) - \psi(\mathbf{x}) \left(\frac{\hbar}{i} \right) \nabla \psi^*(\mathbf{x}) \right] \qquad (8.16)$$

The velocity density—or the probability flux—satisfies the equation of continuity

$$\frac{\partial(\psi \psi^*)}{\partial t} + \nabla \cdot \mathbf{j} = 0 \qquad (8.17)$$

In this way the corpuscular view from classical mechanics is replaced using concepts from continuum mechanics. This follows naturally from the fact that the information about a particle's position is distributed through space. In passing it is noted that both members of the equation of continuity are guaranteed real. The concept of the 'velocity' in these paragraphs needs further elucidation, which will be done below once the dynamics of quantum mechanics has been elucidated.

Other averages that depend on $\mathbf{p}$ can be calculated along the same lines. In particular the kinetic energy $T = p^2/(2m)$ is

$$< T >= -\frac{\hbar^2}{2m} \int d^3 x \, \psi^*(\mathbf{x}) \nabla^2 \psi(\mathbf{x}) \qquad (8.18)$$

These formulas will be useful in a further development of the subject. Summarising then what has been set up so far.

- Experiment suggests that subatomic particles should be described as wave-like objects.
- Describing particles as waves by associating momentum and energy with wavelength and frequency through Planck's constant necessarily implies that the wave function is a complex object.
- Quite unlike classical mechanics a probability interpretation is required and the expectation values of physical quantities can be determined.
- Sharp values of position and momentum cannot be attained simultaneously, but the uncertainty is confined by Heisenberg's uncertainty relation.
- The probability density current satisfies the equation of continuity.

8.3.1 The Spreading of a Wave Packet

Setting $E = p^2/2m$ the time development of a wave packet can be calculated. The wave function that has average speed zero is

$$\psi(x, t) = C \int dp \exp\left(-\frac{p^2}{\Delta p^2}\right) \exp\left(\frac{ipx}{\hbar}\right) \exp\left(-\frac{p^2}{2m\hbar}t\right) \qquad (8.19)$$

Where C is given by the normalising condition

$$\int dx\, \psi(x, t)\psi^*(x, t) = 1 \qquad (8.20)$$

The expectation value of the position is obviously $< x >= 0$, but the width of the distribution $\Delta x^2 =< (x- < x >)^2 >$ may be calculated and is easily obtained as

$$\Delta x^2 = \frac{\hbar^2}{\Delta p^2} + \frac{\Delta p^2 t^2}{4m^2} \qquad (8.21)$$

The packet becomes wider over time; it spreads.

Note that the energy expectation value remains constant and even though the mean velocity of the packet is zero it has a value: $E =< T >= \Delta p^2/8m$.

If there is a mean momentum as well $< p >\neq 0$, then the whole packet moves as a whole—all the while spreading—with velocity $< p > /m$.

8.4 Schrödinger's Equation

The previous section has been entirely concerned with free particles. The question now is how the ideas of wave mechanics can be extended to particles in a potential field. There is no reason why any of the concepts developed so far should remain valid when particles are not free. The best that can be done is to be very clear about what can be assumed to be retained and try to test those assumptions for experimental veracity and mathematical consistency.

The first assumption that shall be made is that Galileo's inertial principle (Newton's first law) remains valid. This assumption concerns a wider notion of Nature, in that the inertial principle is a property of the cosmos and there is no reason to believe that it should not apply to the world of the very small with which quantum mechanics is concerned. More assumptions are necessary. It will be assumed that even in the presence of a potential the state of a physical system can still be described by a wave function $\psi(\mathbf{x}, t)$, or conversely by $\phi(\mathbf{p}, t)$; the two are connected by Fourier transformation in the way described in the previous section. The probability interpretation is also still applicable, though at this stage it is not possible to maintain that the equation of continuity for the probability current density holds. However, for the moment the equation of continuity will be assumed to hold and will be verified afterwards. Expectation values of physical quantities continue to be calculated as developed above. These assumptions are therefore equivalent to saying: the presence of a potential gives form to the wave function, but the conceptual interpretation remains unaltered. The next step is to figure out how the potential defines a wave function. There is no obvious form for a canonical formalism that could be employed, which would be a building block for the verification of some of the assumptions for mathematical consistency.

There are various other considerations. First of all, any equation for the wave function needs to be able to address the fact that the wave function is essentially complex. Secondly, it was noted in the previous section that an equation for the wave function should give the development in space and time, so an equation that describes it ought to include both these elements. The third consideration is more fundamental. A differential equation for either ψ or ψ^* requires boundary conditions. These require that the wave function (and its gradient) vanish at the boundaries of the problem and furthermore that $\int \psi \psi^* d^3 x = 1$, but they are silent about the definition of the actual state of the system, which has to be imposed somehow. The question then is how the state can be defined. It surely has to be via the imposition of some physical quantity (preferably a conserved one). The expectation values of the position and/or the velocity at some time point are a possibility. These, however, are not conserved quantities. Moreover, these quantities are subject to Heisenberg's uncertainty relations and cannot be specified at the same time. The obvious choice for the imposition of the specification of the state of the system is the total energy, which has the additional advantage of being a conserved quantity.

The total energy of the quantum system should be derivable from the wave function. For a free particle the energy satisfies

$$\frac{i}{2} \left(\frac{\partial \psi}{\partial t} \psi^* - \frac{\partial \psi^*}{\partial t} \psi \right) - \frac{E}{\hbar} \psi \psi^* = 0 \qquad (8.22)$$

From which it follows that

$$\frac{i}{2} \int_{t_1}^{t_2} dt \int d^3 x \left(\frac{\partial \psi}{\partial t} \psi^* - \frac{\partial \psi^*}{\partial t} \psi \right) - (t_2 - t_1) \frac{E}{\hbar} = 0 \qquad (8.23)$$

It will now be *assumed* that *the same expression holds for particles in a potential field.*

As quantum theory is basically a continuum theory, no kinetic and potential energy as such can be defined; however, an *energy density* can be defined, which is then the kinetic energy density $\mathcal{T}(\mathbf{x}, t)$ and the potential energy density $\mathcal{V}(\mathbf{x}, t)$. The latter is easily seen to be $\mathcal{V}(\mathbf{x}, t) = V(\mathbf{x}) \psi(\mathbf{x}, t) \psi^*(\mathbf{x}, t)$.

The kinetic energy density requires a little work. The expectation value of the kinetic energy for free particles is given by Eq. (8.18). This gives a clue as to what is possible. Note that the integrand in (8.18) may be rewritten as (time-dependence is implied)

$$\psi^*(\mathbf{x})\nabla^2\psi(\mathbf{x}) = \nabla \cdot \left(\psi^*(\mathbf{x})\nabla\psi(\mathbf{x})\right) - \nabla\psi^*(\mathbf{x}) \cdot \nabla\psi(\mathbf{x}) \tag{8.24}$$

Assuming that both ψ and $\nabla\psi$ vanish at infinity the integral over the first term vanishes. The expectation value of the kinetic energy for free particles is then

$$< T >= \frac{\hbar^2}{2m} \int d^3x \nabla\psi^*(\mathbf{x}) \cdot \nabla\psi(\mathbf{x}) \tag{8.25}$$

The same holds for the complex conjugate. For particles in a potential field the expectation value of the kinetic energy need not be the same, of course. However, energy measures are isotropic and on dimensional grounds the following form for the kinetic energy density may be put forward with three coefficients a, b and c

$$\mathcal{T} = a\left[\psi^*(\mathbf{x})\nabla^2\psi(\mathbf{x}) + \psi(\mathbf{x})\nabla^2\psi^*(\mathbf{x})\right]$$
$$+ b\left[\psi^*(\mathbf{x},t)\nabla^2\psi(\mathbf{x},t) - \psi(\mathbf{x},t)\nabla^2\psi^*(\mathbf{x},t)\right] + c\nabla\psi(\mathbf{x},t) \cdot \nabla\psi^*(\mathbf{x},t) \tag{8.26}$$

This is recast in the form

$$\mathcal{T} = a\nabla \cdot \left[\psi^*(\mathbf{x})\nabla\psi(\mathbf{x}) + \psi(\mathbf{x})\nabla\psi^*(\mathbf{x})\right]$$
$$+ b\nabla \cdot \left[\psi^*(\mathbf{x},t)\nabla\psi(\mathbf{x},t) - \psi(\mathbf{x},t)\nabla\psi^*(\mathbf{x},t)\right] + (c-2a)\nabla\psi(\mathbf{x},t)\nabla\psi^*(\mathbf{x},t) \tag{8.27}$$

The term that is proportional to b is recognised as the divergence of the probability current density and therefore it satisfies the equation of continuity (8.17), which gives rise to $\dot{\psi}\psi^* + \dot{\psi}^*\psi$. Requiring the kinetic energy density on integrating over space to produce the expectation value for free particles gives

$$c - 2a = \frac{\hbar^2}{2m} \tag{8.28}$$

The total energy is

$$E = \frac{1}{t_2 - t_1} \int_{t_1}^{t_2} dt \int d^3x (\mathcal{T} + \mathcal{V}) \tag{8.29}$$

As E is a constant, a variation in this expression is stationary, which is enforced with the condition (8.23). So, with a Lagrange multiplier λ the integral expression that has to be made stationary is

$$\int_{t_1}^{t_2} dt \int d^3x (\mathcal{T} + \mathcal{V}) + \lambda\frac{i}{2} \int_{t_1}^{t_2} dt \int d^3x \left(\frac{\partial\psi}{\partial t}\psi^* - \frac{\partial\psi^*}{\partial t}\psi\right) \tag{8.30}$$

The two integrals are combined and then the integrand is a functional $\mathcal{F}$

$$\mathcal{F} = a\left[\psi^*\nabla^2\psi + \psi\nabla^2\psi^*\right)\right] + c\nabla\psi\cdot\nabla\psi^* + b\left[\psi^*\nabla^2\psi - \psi\nabla^2\psi^*\right] +$$
$$+V\psi\psi^* + \lambda\frac{i}{2}\left(\frac{\partial\psi}{\partial t}\psi^* - \frac{\partial\psi^*}{\partial t}\psi\right) \qquad (8.31)$$

The canonical variables are ψ and ψ^* (there are two, because ψ is a complex number). The Euler equation associated with ψ^* is

$$\frac{\partial\mathcal{F}}{\partial\psi^*} - \nabla\cdot\left(\frac{\partial\mathcal{F}}{\partial\nabla\psi^*}\right) - \frac{\partial}{\partial t}\left(\frac{\partial\mathcal{F}}{\partial\dot\psi^*}\right) + \nabla^2\left(\frac{\partial\mathcal{F}}{\partial(\nabla^2\psi^*)}\right) = 0 \qquad (8.32)$$

And that leads to

$$V\psi + (2a - c)\nabla^2\psi + i\lambda\frac{\partial\psi}{\partial t} = 0 \qquad (8.33)$$

Doing the same for the Euler equation associated with ψ then produces

$$V\psi^* + (2a - c)\nabla^2\psi^* - i\lambda\frac{\partial\psi^*}{\partial t} = 0 \qquad (8.34)$$

To determine what λ is Eq. (8.33) is multiplied by ψ^* and Eq. (8.34) by ψ. Adding up and using (8.22) leads to

$$2V\psi\psi^* + (2a - c)(\nabla^2\psi\psi^* + \nabla^2\psi^*\psi) + \frac{2E\lambda}{\hbar}\psi\psi^* = 0 \qquad (8.35)$$

The total energy must have a term that equals the potential energy and therefore $\lambda = -\hbar$. Taking it all together, the equation for the wave function is *Schrödinger's equation*

$$-\frac{\hbar^2}{2m}\nabla^2\psi(\mathbf{x}, t) + V\psi(\mathbf{x}, t) = i\hbar\frac{\partial\psi(\mathbf{x}, t)}{\partial t} \qquad (8.36)$$

What needs to be done now is to calculate the consequences of this equation and then compare the results with experiments. If that comparison comes out positive, then it is plausible that the assumptions made here are correct.

8.5 The Schrödinger Equation and the Hamilton-Jacobi Equation

In the previous section a number of assumptions have been made. These are essentially expressing the idea that the framework put forward for free particles remains valid for particles in a potential field. Quite a different assumption can be made, however. Following the correspondence found for the propagation of light with the Hamilton-Jacobi equations, an entirely classical analysis can be carried out for material waves. To that end the action S is embedded in a wave function, just as was done for light, see Sect. 5.5.1

$$\psi(\mathbf{r}, t) = \psi_0 \exp\left[\frac{i}{\hbar} S(\mathbf{r}, t)\right] \tag{8.37}$$

The Hamilton-Jacobi equations for a particle in a potential field V are

$$\frac{\partial S}{\partial t} + \frac{1}{2m}\left(\frac{\partial S}{\partial r_i}\right)^2 + V(\mathbf{r}, t) = 0 \tag{8.38}$$

From the ansatz (8.37) it follows that

$$\frac{\partial S}{\partial r_j} = -\frac{i\hbar}{\psi}\frac{\partial \psi}{\partial r_j}, \quad \frac{\partial S}{\partial t} = -\frac{i\hbar}{\psi}\frac{\partial \psi}{\partial t} \tag{8.39}$$

Differentiating

$$\frac{\partial^2 \psi}{\partial r_j^2} = \frac{i}{\hbar}\frac{\partial \psi}{\partial r_j}\frac{\partial S}{\partial r_j} + \frac{i}{\hbar}\psi\frac{\partial^2 S}{\partial r_j^2} \tag{8.40}$$

In the next step it is emphasised that this is a strictly *classical analysis*. The momentum $\mathbf{p} = m\mathbf{v} = m\dot{\mathbf{r}}$ follows from the partial differentiation of the action, so

$$\frac{\partial^2 S}{\partial r_j^2} = \frac{\partial}{\partial r_j} p_j = \frac{\partial}{\partial r_j}\frac{dr_j}{dt} = \frac{d}{dt}\frac{\partial r_j}{\partial r_j} = 0 \tag{8.41}$$

Now, using the Hamilton-Jacobi equations (8.40) is easily evolved to the Schrödinger equation

$$-\frac{\hbar^2}{2m}\nabla^2\psi + V(\mathbf{r}, t) = i\hbar\frac{\partial \psi}{\partial t} \tag{8.42}$$

How is it possible that an essentially classical analysis produces the correct quantum description? In part it is the universality of the Hamilton-Jacobi equations, that links to the energy as a conserved quantity. Otherwise, it should be pointed out that there is something wrong with this analysis, correct as it may appear from a classical point of view. The problem comes when the identification is made of the velocity as a time derivative of the displacement. Basically what has happened in Eq. (8.41) is that the *expectation value* of the momentum has been inserted. Sakurai (Sakurai 1985) writes 'We would like to caution the reader against a too literal interpretation of [the probability current density] $\mathbf{j}$ as ρ times the velocity defined at every point in space, because a simultaneous precision measurement of position and velocity would necessarily violate the uncertainty principle'. The same issue comes to the fore in the derivation of the equation of continuity, Eq. (8.17).

What then is the status of the Hamilton-Jacobi equation in quantum mechanics? Starting from the Schrödinger equation and setting $\psi = \exp(i S/\hbar)$ it is easily inferred that

$$\frac{\partial S}{\partial t} + \frac{(\nabla S)^2}{2m} - \frac{i\hbar}{2m}\nabla^2 S + V(\mathbf{r}, t) = 0 \tag{8.43}$$

This is almost the Hamilton-Jacobi equation, but for the term proportional to $\hbar$. In the limit $\hbar \to 0$, which is the *classical limit*, it is exactly the classical equation. Quantum effects are inserted by the presence of the $\hbar$, the i—making it a fundamentally complex theory—and the $\nabla^2 S$, which would be zero in the classical approach according to Eq. (8.41) where the momentum is replaced by its expectation value.

The substitution of an action in the wave function can be made somewhat more general (see Landau (1976)). Set for the wave function $\psi = A\exp(iS/\hbar)$, where both A and S are supposed to be real. The Schrödinger equation then takes the form

$$A\frac{\partial S}{\partial t} - i\hbar\frac{\partial A}{\partial t} + \frac{A}{2m}(\nabla S)^2 - \frac{i\hbar}{2m}A\nabla^2 S - \frac{i\hbar}{m}\nabla A \nabla S - \frac{\hbar^2}{2m}\nabla^2 A + VA = 0 \tag{8.44}$$

Evaluating the real and imginary parts of this gives the following equations

$$\frac{\partial S}{\partial t} + \frac{1}{2m}(\nabla S)^2 - \frac{\hbar^2}{2ma}\nabla^2 A + V = 0 \tag{8.45}$$

$$\frac{\partial A}{\partial t} + \frac{1}{2m}A\nabla^2 S + \frac{1}{m}\nabla A \nabla S = 0 \tag{8.46}$$

Taking the classical limit $\hbar \to 0$ in the first of these equations is again the Hamilton-Jacobi equation, as expected. Multiplying the second equation with $2A$ yields

$$\frac{\partial A^2}{\partial t} + \operatorname{div}\left(A^2\frac{\nabla S}{m}\right) = 0 \tag{8.47}$$

Now, A^2 is the probability density, while $\nabla S/m$ is the 'classical velocity' $\mathbf{v}$. Therefore, the equation represents the classical continuity equation for the probability density, just like Eq. (8.17).

8.6 Solutions of the Schrödinger Equation

Once the Schrödinger equation is available, it is a matter of supplying the potential V and a solution for the wave function can be obtained. That, however, is a slight oversimplification of affairs. As an illustration of the problems that arise an elementary form of the potential is assumed and—with the supply of suitable boundary conditions—a set of solutions will be constructed.

First the *time-independent Schrödinger equation* is developed. The Schrödinger equation is Eq. (8.36)

$$-\frac{\hbar^2}{2m}\nabla^2\psi(\mathbf{x},t)+V(\mathbf{x})\psi(\mathbf{x},t)=i\hbar\frac{\partial\psi(\mathbf{x},t)}{\partial t} \tag{8.48}$$

The time and space part of the solution may be separated

$$\psi(\mathbf{x},t)=\psi(\mathbf{x})\exp\left(-\frac{iEt}{\hbar}\right)\ \rightarrow\ -\frac{\hbar^2}{2m}\nabla^2\psi(\mathbf{x})+V(\mathbf{x})\psi(\mathbf{x})=E\psi(\mathbf{x}) \tag{8.49}$$

The result is called the *time-independent Schrödinger equation* (sometimes simply *the* Schrödinger equation) and it depends on the position only. Mathematically speaking Eq. (8.49) has the form of an eigen value problem. The eigen vectors are the functions $\psi_E(\mathbf{x})$ that belong to the eigen value E. There is no reason to assume that there is only one solution; there may be a range of eigen values and eigen vectors for a given potential form. E is easily identified as an energy and the set of values of E for a given potential form is called the *energy spectrum*. Once they have been determined the time-dependent solution may be obtained from the initial conditions, if they can be expressed in the eigen vectors. For each eigen vector at time $t=0$ $\psi_E(\mathbf{x},0)$ the full time-dependent solution is obtained by

$$\psi(\mathbf{x},t)=\psi_E(\mathbf{x},0)\exp\left(-\frac{iEt}{\hbar}\right) \tag{8.50}$$

Superposition makes it possible to give the time-dependence of an arbitrary initial wave function, as long as it is decomposed into eigen states. This is clearly only possible when the eigen states form a *complete set*, which would imply that the eigen functions are orthonormal. The completeness of the problem will here be assumed (the property of completeness can sometimes be proven quite easily, but in other cases in quantum mechanics the proof cannot be given and one must proceed with the assumption). Writing the initial state as the sum of eigen states with coefficients c_n

$$\psi(\mathbf{x},0)=\sum_n c_n\psi_n(\mathbf{x})\ \rightarrow\ \psi(\mathbf{x},t)=\sum_n c_n\psi_n(\mathbf{x})\exp\left(-\frac{iE_nt}{\hbar}\right) \tag{8.51}$$

8.6.1 The Infinite Potential Well

Nearly all introductory books on quantum mechanics (Bes 2013; Gasiorowicz 1995; Goswami 1992; Landau 1976; McMurry 1994; Merzbacher 1998; Sakurai 1985) treat the problem of the one-dimensional infinite potential well. While physically it is at best an idealisation, mathematically this is quite an easy problem and the development of it points to properties of other quantum systems.

The potential has the form

$$V(x) = 0 \text{ for } -L \leq x \leq L \text{ and } V(x) \to \infty \text{ for } |x| > L \qquad (8.52)$$

The time independent Schrödinger equation is Eq. (8.49)

$$-\frac{\hbar^2}{2m}\frac{\partial^2 \psi(x)}{\partial x^2} = E\psi(x) \text{ for } -L \leq x \leq L \qquad (8.53)$$

which has symmetric (s) and antisymmetric (a) solutions

$$\psi(x) = A_s \sin \lambda_s x \text{ and } \psi(x) = A_a \cos \lambda_a x \text{ with } \lambda_{s,a} = \frac{\sqrt{2mE_{s,a}}}{\hbar} \qquad (8.54)$$

The imposition of boundary conditions at $x = \pm L$ yields the values for $E_{s,a}$. The conditions are that the wave function must vanish where the potential is infinite, as the probability of finding the particle in such regions would imply an infinite energy, which is physically not possible. It follows that

$$E_s = \frac{1}{2m}\left(\frac{2n+1}{2}\frac{\pi\hbar}{L}\right)^2 \quad n = 0, 1, 2... \quad ; \quad E_a = \frac{1}{2m}\left(\frac{\pi\hbar k}{L}\right)^2 \quad k = 1, 2...$$
$$(8.55)$$

Together they make the energy spectrum. It is seen that the lowest energy state—the *ground state*—is obtained for $n = 0$. The next level up—the first *excited state*—is for $k = 1$. And so on, subsequent levels alternate between symmetric and asymmetric modes. The eigen functions are recognised as a Fourier set; they are clearly a complete set.

The whole energy spectrum may be combined as

$$E_n = \frac{1}{2m}\left(\frac{n}{2}\frac{\pi\hbar}{L}\right)^2 \quad n = 0, 1, 2... \qquad (8.56)$$

8.6.1.1 Periodic Boundary Conditions

Rather than work with symmetric and antisymmetric modes the system can be approached by imagining that the system is replicated numerous times. In that case the wave function is invariant under a translation of the coordinate by an amount $\pm 2L$, in other words

$$\psi(x) = \psi(x \pm 2L) \text{ or } \exp(i\lambda x) = \exp(i\lambda x \pm 2i\lambda L)$$
$$\to \exp(\pm 2i\lambda L) = 1 \to \lambda = \frac{n\pi}{2L}; \quad n = 0, 1, 2... \qquad (8.57)$$

This produces the whole energy spectrum directly without needing to distinguish between symmetric and asymmetric modes.

8.6.1.2   Three-Dimensional Infinite Potential Well

Extending to the three-dimensional case is easy; it describes a particle in a cubic box with side a and volume a^3. The problem is separable, so the solution of the Schrödinger equation is simply the product of three one-dimensional solutions. The energy levels are denoted by three quantum numbers

$$E_{n_1,n_2,n_3} = \frac{\epsilon_0}{3}(n_1^2 + n_2^2 + n_3^2) \text{ with } \epsilon_0 = \frac{3\pi^2\hbar^2}{2a^2m} \qquad (8.58)$$

Now set $n^2 = n_1^2 + n_2^2 + n_3^2$ the number of energy levels dN between E and $E + dE$ is

$$dN = \frac{1}{8}4\pi n^2 dn = \frac{m^{3/2}\sqrt{E}a^3}{\sqrt{2}\pi^2\hbar^3}dE \qquad (8.59)$$

This is obviously only valid when the levels near the ground level are not occupied. In that case a bit of statistical mechanics can be done for a box at temperature T, see Sect. 7.6. The one-particle partition function is ($\beta = 1/kT$)

$$\tilde{Z} = \tilde{K}\sum\exp(-\beta E_n) \rightarrow \tilde{Z} = \tilde{K}\frac{m^{3/2}V}{\sqrt{2}\pi^2\hbar^3}\int_0^\infty \sqrt{E}dE = \frac{\tilde{K}m^{3/2}V}{2\sqrt{2}\pi^{3/2}\hbar^3\beta^{3/2}}$$
$$(8.60)$$

The internal energy and the pressure of this system are simply

$$\tilde{U} = -\frac{\log\partial\tilde{Z}}{\partial\beta} = \frac{3}{2}kT; \quad p = -\frac{\partial\tilde{F}}{\partial V} = \frac{1}{\beta}\frac{\partial\log\tilde{Z}}{\partial V} = \frac{kT}{V} \qquad (8.61)$$

For N particles the results of the extensive quantities are multiplied by N. It is noted that these quantum relations yield the classical perfect gas, which was expected in some sense, but good to have verified.

When the lower levels are occupied as well—a situation that applies to very low temperature—then this approach clearly does not work and the discrete nature of the energy levels has to be accounted for. Realistically, it must be acknowledged that at very low temperature other effects begin to play a role. These other effects are concerned with the fact that the statistics of quantum particles is different than the statistics of classical particles as used in Sect. 7.3. The reason is that quantum particles are indistinguishable, as opposed to classical particles that have an individual identity, or as Einstein pithily summarised it: *man can ein Elektron nicht rot anstreichen* (one cannot mark an electron with a red pencil). The other matter to be addressed is that for a certain class of quantum particles, so-called *fermions*, the *exclusion principle* applies. This principle, ascribed to Pauli, holds that each energy level can only accommodate one such a particle. There is nothing like it in classical mechanics. Examples of fermions are electrons, protons and neutrons, which are all particle types that are commonly encountered in atomic physics. The other class is called *bosons* and for these there is no limit on the number of particles each energy level can accommodate. Examples of these are α-particles ($^4_2$He) and pairs of electrons that operate together under certain circumstances in a solid (so-called Cooper pairs). The implication for bosons is that at very low temperature the vast majority of available particles occupy the ground energy level and when that happens the system acquires peculiar properties such as *superconductivity* and *superfluidity*, see Schmitt (2014). In the former electric currents can flow with no Ohmic resistance and a superfluid has zero viscosity.

The theory of quantum statistics requires a much more extensive discussion of the intrinsic properties of quantum particles and is outside the scope of this text. For a start the Boltzmannian approach, as shown in Sect. 7.3 is no longer valid, as it is implicitly assumed there that the particles are distinguishable. The analysis relevant to Bose-Einstein and Fermi-Dirac statistics is extensively discussed in the statistical mechanics literature, see for example, Ter Haar (1966).

Whether particles are bosons or fermions is associated with the intrinsic particle property of spin. Below in Sect. 8.9.5 the elements of this property are discussed. This aspect of quantum mechanics strays very far from classical mechanics and therefore there is no direct limit that can be taken that leads back to Newtonian concepts.

8.6.2 The Hydrogen Atom

One of the most far-reaching applications of the Schrödinger equation is the description of the hydrogen atom. It is the most abundant atom of 'ordinary' (that is non-'dark') matter in the Universe, accounting for some 75 percent of it. For a physics student therefore, the calculation of the hydrogen atom is a significant achievement, as it opens up a lot of what can be understood in atomic physics and astronomy. Moreover, the calculation can be done entirely analytically, unlike the theoretical analysis of other atoms (these are many-body problems and approximations have to be made to solve even the next simplest element, helium). Because of its importance, the solution of hydrogen is described in most books on quantum mechanics and texts on physical chemistry (Hofmann 2018; Moelwyn-Hughes 1966).

The hydrogen atom consists of a proton and an electron. These have opposite and equal charges, so they attract. The question is whether it can be assumed that the Coulomb interaction still holds at the sub-atomic scale. There is only one way of finding that out: try it and compare the results with experiment. More in particular, the energy levels of the hydrogen atom are calculated using Schrödinger's equation and the spectrum of light that is obtained when the energy state of the atoms changes is measured. It transpires that Coulomb's law is still valid at the sub-atomic level. The mass ratio of the proton to the electron is about 1800 and therefore it is a very reasonable approximation to regard the proton as 'standing still', so what is calculated is a particle in a Coulomb central potential field; its mass is m_0 and its charge is e. The potential is $V(r) = -4\pi e^2/r$ (in Gaussian units). It is convenient to use polar coordinates r, θ, ϕ and to write the Laplacian operator in the time-independent Schrödinger's equation in these coordinates

$$-\frac{\hbar^2}{2m_0}\left(\frac{1}{r}\frac{\partial^2}{\partial r^2}(r\psi)+\frac{1}{r^2\sin\theta}\frac{\partial}{\partial\theta}\left(\sin\theta\frac{\partial\psi}{\partial\theta}\right)+\frac{1}{r^2\sin^2\theta}\frac{\partial^2\psi}{\partial\phi^2}\right)-\left(\frac{e^2}{4\pi r}+E\right)\psi=0$$

$$(8.62)$$

A separable solution is tried. Set $\psi(r,\theta,\phi)=U(r)P(\theta,\phi)$ then, with a constant λ,

$$-\frac{\hbar^2}{2m_0}\frac{1}{r}\frac{\partial^2}{\partial r^2}(rU)-\left(\frac{e^2}{4\pi r}+E\right)U=\frac{\lambda}{r^2}U \qquad (8.63)$$

$$-\frac{\hbar^2}{2m_0}\left[\frac{1}{\sin\theta}\frac{\partial}{\partial\theta}\left(\sin\theta\frac{\partial P}{\partial\theta}\right)+\frac{1}{\sin^2\theta}\frac{\partial^2 P}{\partial\phi^2}\right]=\lambda P \qquad (8.64)$$

The analogous problem in classical mechanics has demonstrated that a solution of the orbits is defined by specifying the energy and two components of the angular momentum. Quantum mechanics is no different, except that the energy is quantised and, similarly, *the angular momentum is also quantised*. This should not come as a surprise: in the three-dimensional potential well three quantum numbers needed to be specified. The angular momentum is entirely defined by Eq. (8.64) and the energy values by (8.63). The latter is achieved by enforcing the boundary conditions, just as in the case of the infinite potential well. There are two boundary conditions, one states that for $r \to \infty$ the solution must remain finite, reflecting the fact that *bound states* are sought (these have negative energy) and the other that for $r \to 0$ there must be no singularity. A singularity would signal an infinite probability, which is clearly not possible.

The solution can be approached by realising that the problem is invariant around the origin. That implies that under a coordinate rotation the solution either results in itself, or that there are groups of solutions that result in themselves. The angular solutions must therefore be periodic. These requirements are achieved by looking at a class of solutions of (8.64) known as spherical harmonics. These are characterised by two integer numbers l and m and are denoted by Y_{lm}, so $P(\theta,\phi)=Y_{lm}(\theta,\phi)$. The requirement that the function $P(\theta,\phi)$ be single-valued implies that $\lambda=-2m_0l(l+1)/\hbar^2$, where l is an integer and that $|m| \le l$. The mathematical details of these functions (which appear in many physical problems with spherical symmetry) are in books (Boas 1983; Jackson 1962; Merzbacher 1998).

The spherical harmonics have the property of orthonormality; they are essentially the product of associated Legendre polynomials $P_l^m(cos\theta)$ and $\exp(im\phi)$. (for an analysis of associated Legendre polynomials see Abramowitz and Stegun (1972), Boas (1983) and Merzbacher (1998)).

A convenient scaling can be implemented: $\rho=m_0e^2/(2\pi\hbar^2)r$ and $\mathcal{E}=8\pi^2\hbar^2E/(m_0e^4)$; the radial equation takes the form

$$\frac{1}{\rho}\frac{\partial^2 U}{\partial\rho^2}+2\frac{1}{\rho}\frac{\partial U}{\partial\rho}+\left(\frac{1}{\rho}+\mathcal{E}\right)U-\frac{l(l+1)}{\rho^2}U=0 \qquad (8.65)$$

The behaviour for $\rho \rightarrow \infty$ is important for the enforcement of the boundary conditions. In this limit the radial equation takes the form

$$\frac{\partial^2 U}{\partial \rho^2} + \mathcal{E}U = 0 \quad \rightarrow \quad U(\rho) \approx \exp\left(\pm\sqrt{-\mathcal{E}}\rho\right) \tag{8.66}$$

To satisfy the boundary conditions the minus sign must be chosen; it is also observed that localised (bound) states are achieved if $\mathcal{E} < 0$, as expected.

The radial equation has a singularity at $\rho \rightarrow 0$, which can be remedied by setting $U(\rho) = \rho^s f(\rho) \exp\left(-\sqrt{-\mathcal{E}}\rho\right)$, where $f(\rho)$ is a function that is finite everywhere and satisfies

$$\rho^2 \frac{\partial^2 f}{\partial \rho^2} + \frac{\partial f}{\partial \rho}\left(2\rho(s+1) - 2\rho^2\sqrt{-\mathcal{E}}\right) + f\left(s + s^2 - 2\rho s\sqrt{-\mathcal{E}} + \rho - l(l+1)\right) \tag{8.67}$$

Now take the limit $\rho \rightarrow 0$

$$s(s+1) - l(l+1) = 0 \quad \rightarrow \quad s = l \text{ and } s = -l - 1 \tag{8.68}$$

Requiring that the solution is not singular in the origin leaves $s = l$. The solution for $f(\rho)$ is obtained by trying a power series

$$f(\rho) = \sum_{i=0}^{\infty} a_i \rho^i \tag{8.69}$$

Substitution and requiring that each term proportional to ρ^{i+1} vanishes gives the following recursion

$$[(i+l+1)\sqrt{-\mathcal{E}} + 1]a_i + (i+1)(i+2l+2)a_{i+1} = 0 \tag{8.70}$$

For large i the coefficients go up as $1/i$, which, if taken to infinity, would destroy the exponential character of the solution and therefore the compliance with the boundary conditions. It follows that the series must be broken off by requiring that for some i the coefficient vanishes. In the recursion (8.70) that is achieved by ensuring that the term proportional to a_i vanishes. The condition for the scaled energy $\mathcal{E}$ is then

$$\mathcal{E} = -\frac{1}{4(i+l+1)^2} \tag{8.71}$$

The energy levels so obtained are numbered $n = 1, 2, 3...$ and labelled $\mathcal{E}_n$, hence, $n = i + l + 1$; now, because $i \geq 0$ it follows that $n - l - 1 \geq 0$, or $l \leq n - 1$. Each energy level, other than the ground level, really corresponds to a number of states with different numbers for l and m. For example, the third level can be manifest by the combinations n, l, m of $3, 2, \pm 2$, $3, 2, \pm 1$, $3, 2, 0$, $3, 1, \pm 1$, $3, 1, 0$ and $3, 0, 0$.

In total for each n there are n^2 states. The multiplicity of the states goes under the name of *degeneracy*.

The ground state, $n = 1$, has a wave function $\psi_1 = A \exp{-\sqrt{-\mathcal{E}_1}\rho}$. Note that this is completely spherical and therefore entirely different from the idea of a classical orbit of a body moving around. Its energy is $E_1 = -m_0 e^4/(8\pi^2 \hbar^2)$. In atomic physics energies are usually expressed in electron volts and $E_1 = -13.6 eV$. This is minus the minimum energy required to strip the hydrogen atom of its electron and is known as the *ionisation energy*.

The probability of finding the electron in a shell of thickness $d\rho$ at the distance ρ is $4\pi \psi(\rho)\psi^*(\rho)\rho^2 d\rho$. The maximum is at $\rho_0 = 1/\sqrt{-\mathcal{E}_1}$. In unscaled units ρ_0 corresponds to the so-called *Bohr radius*, r_0, and is $r_0 = 5.29 \times 10^{-11}m$ or some 53 Angstrom. All these numbers have been extensively confirmed by experiment, as has the spectrum that is associated with 'jumps' in energy between states. The degeneracy can be made visible by breaking the symmetry of the spherical potential, for example, by applying an electric field. The spectrum then acquires a *fine structure*, which can be calculated using perturbation theory and its experimental verification represents surely a triumph of quantum mechanics. The fine structure associated with an electric field is known as the *Stark effect*. In a further development the spectrum may exhibit an extra fine structure that does not follow from the Schrödinger equation with a potential field. This fine structure, which is exhibited when a magnetic field is applied, is associated with a *quantised intrinsic angular momentum* of the electron (and of many other elementary particles, such protons and neutrons) known as *particle spin*, see Sect. 8.9.5.

How does the quantum solution converge to the classical solution for a body in a central $1/r$ potential field? As observed, the ground state, $n = 1$, does not look in the slightest like a classical planetary orbit. For the potential well it was discovered that the classical features appear for the higher quantum numbers and that distinctive quantum properties typically arise when the quantum numbers are near the ground state (at low temperature). Therefore, an investigation at high numbers should reveal the classical features also for the hydrogen atom. In the literature one frequently finds arguments for a classical analogy by studying transitions from one energy level to the next. The electron can change energy state and the difference ΔE is emitted as a photon with circular frequency $\Delta E/\hbar$. The classical case would consist of radiation associated with the electron going in a circular orbit. The loss in orbital energy must equal the energy of the radiation. Indeed, for a transition that changes the n quantum number by an amount of $\Delta n = 1$ the calculation of the classical radiation corresponds well with the quantum emission in the limit $n \to \infty$. However for $\Delta n = 2$ it does not work. Now, account must be taken of the fact that such a transition is not possible from a quantum point of view. The reason is that there are *selection rules*. The photon has an intrinsic angular momentum ('spin') and in a transition the total angular momentum must be conserved, which then leads to the fact that not every transition is possible. So, the argument that the quantum result is equivalent to the classical result for large n necessarily needs to invoke quantum mechanics itself: the spin of the photon. This weakens the case.

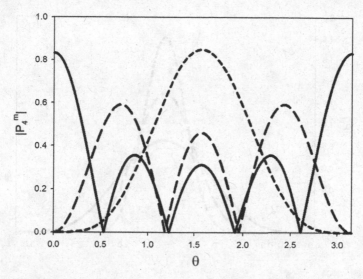

Fig. 8.1 The modulus of the normalised associated Legendre functions: solid line $|P_4^0|$, long dash $|P_4^2|$, short dash $|P_4^4|$

There is a way of arriving at classical properties for large quantum numbers that does not require the transition. It is in the properties of the orbits themselves. To see how this works the angular part of the wave function is investigated. Interest is focussed on the absolute value of the associated Legendre functions $P_l^m(\cos\theta)\exp(im\phi)$. It is immediately seen that this does not depend on ϕ, so that only the θ dependence needs to be examined. As an example consider $|P_4^m|$ for $m = 0, 2, 4$ (it is not necessary to investigate negative m, as these give the same answer because the modulus is evaluated only).

The functions are depicted in Fig. 8.1 and have been normalised to unity for easy comparison. It is seen that as m increases to its maximum value—in this case $m = 4$—the state becomes more localised in the plane around $\theta = \pi/2$. Now, the angular momentum of the 'electron cloud' is as follows $L^2 = \hbar^2 l(l + 1)$ and the z-component is $L_z = m\hbar$. In other words, the 'orbit' is most concentrated around $\theta = \pi/2$ when the z-component of the angular momentum is maximal. When the value is less than its maximum, the probability of finding the electron is more scattered. And when all angular momentum is concentrated in the z-component the orbit is flattest. That makes sense.

In a further investigation it is examined what happens when m has its maximum value while l is pushed up. In other words examine P_l^l for higher values of l. In Fig. 8.2 the normalised associated Legendre functions are plotted for higher values of l and the maximal value of m. It is found that as the quantum number l increases the orbit becomes more concentrated in the plane at $\theta = \pi/2$ and therefore it becomes flatter and resembles the planetary orbit to a greater extent. This could be said to be

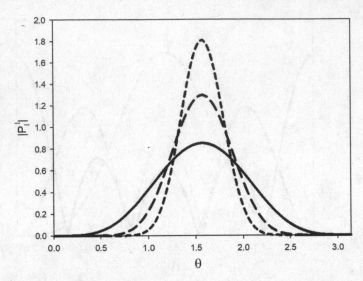

Fig. 8.2 The modulus of the normalised associated Legendre functions: solid line $|P_4^4|$, long dash $|P_{10}^{10}|$, short dash $|P_{20}^{20}|$

the quantum analogon of Kepler's first law and it is the first step towards a classical understanding of the quantum result.

The question can be asked as to what happens with the radial dependence of the wave function. Noting that $n = i + l + 1$ and that the maximum value of l is $n - 1$, it follows that the special case of $i = 0$ is relevant for a high quantum number. At $i = 0$ the radial part of the wave function is $\psi(\rho) = A\rho^{n-1} \exp\left(-\sqrt{-\mathcal{E}_n}\rho\right) = A\rho^{n-1} \exp\left(-1/(2n)\rho\right)$. The probability to find the electron in a shell of thickness $d\rho$ is then is then $\psi\psi^* = 4\pi A^2 \rho^{2n} \exp\left(-\rho/n\right)$, which has a maximum in $\rho = n^2$. It is furthermore easy to calculate the relative deviation $\langle(\rho - <\rho>)^2\rangle/\langle\rho\rangle^2$; this turns out to be always equal to $1/2$, independent of n. Therefore, the classical idea that the electron becomes localised in a narrow orbit does not apply. However, the mean value of the radial dependence is $\langle r \rangle = ca_0[3n^2 - l(l+1)]$, where c is a non-dimensional constant that depends on the electrical units used and a_0 the Bohr radius. Important here is that $\langle r \rangle \propto n^2$ for large n.

Kepler's second law can obviously not be verified, as there is no time dependence in the stationary solutions. Kepler's third law, however, is open to investigation. In classical terms it reads $\langle r \rangle^3 / T_p^2 = cnst$. For circular orbits the classical angular momentum, L_c is $L_c = 2\pi \langle r \rangle^2 / T_p$ and so the third law may be rephrased as $L_c^2 / \langle r \rangle = cnst$. In the quantum analogon $L_c^2 \to \hbar^2 l(l+1)$ and the question now is whether $\hbar^2 l(l+1) \propto \langle r \rangle$ for large n. This is indeed the case as follows from $\langle r \rangle \propto n^2$ and therefore Kepler's third law is recognisable in the quantum mechanical result for large quantum numbers. It does not hold for small quantum numbers; the ground state, for example, has $l = 0$ and $\langle r \rangle = a_0$. Here, again, it is demonstrated that quantum effects deviate most from classical mechanics when quantum numbers are low.

8.6.3 The One-Dimensional Harmonic Oscillator

In the chapters on classical mechanics it was seen that the harmonic oscillator is one of the easiest mechanical problems, see Sect. 1.4.3.1. As it so happens, the classical limit for the quantum mechanical harmonic oscillator is particularly elegant. In addition, the harmonic oscillator turns out to have some mathematical properties that lead to other phenomena in quantum mechanics and, indeed, in quantum field theory. For these reasons this aspect of quantum mechanics is well worth exploring.

The Schrödinger equation for the harmonic oscillator is obtained by setting for the potential $V(x) = 1/2m\omega^2 x^2$

$$-\frac{\hbar^2}{2m}\frac{d^2\psi(x)}{dx^2} + \left(\frac{1}{2}m\omega^2 x^2 - E\right)\psi(x) = 0 \quad . \tag{8.72}$$

A similar procedure is followed as the one pursued for the hydrogen atom. First asymptotic behaviour is established. For large $|x|$ the equation reads

$$-\frac{\hbar^2}{2m}\frac{d^2\psi(x)}{dx^2} + \frac{1}{2}m\omega^2 x^2 \psi(x) = 0 \tag{8.73}$$

which has the solution, again for large $|x|$, $\psi(x) \to A\exp\left(-m\omega/(2\hbar)x^2\right)$. The solution of the Schrödinger equation is derived from the asymptotic solution as

$$\psi(x) = A\sum_n a_n x^n \exp\left(-\frac{m\omega}{2\hbar}x^2\right) \tag{8.74}$$

which leads to the recursion

$$a_{n+2} = -2a_n \frac{E - (n + \frac{1}{2})}{\hbar^2(n+2)(n-1)} \tag{8.75}$$

In order not to destroy the character of the asymptotic solution the recursion must end for a certain n, which is achieved by choosing

$$E_n = \left(n + \frac{1}{2}\right)\hbar\omega; \quad n = 0, 1, 2... \tag{8.76}$$

Two interesting features are seen at once. First, there is a *zero point energy* of $\hbar\omega/2$, implying that the energy is never zero, but there is always a 'vacuum jitter'. Second, the energy levels are equidistant with a gap of $\hbar\omega$ between them.

The recursion (8.75) is a property of Hermite polynomials $\mathcal{H}$. It is convenient now to rescale the position x to $\xi = x\sqrt{m\omega/\hbar}$; the wave function takes the form

$$\psi_n(\xi) = A_n \mathcal{H}_n(\xi)\exp\left(-\xi^2/2\right) \tag{8.77}$$

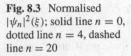

Fig. 8.3 Normalised $|\psi_n|^2(\xi)$; solid line $n = 0$, dotted line $n = 4$, dashed line $n = 20$

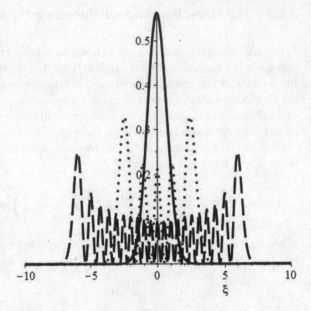

The probability $p_n(\xi) \propto \psi(\xi)\psi(\xi)^*$ is normalised and plotted in Fig. 8.3 for various values of n. Remembering the classical probability as derived in Sect. 1.4.3.1, it is found that as n increases the probability begins to resemble the classical result. Again, it is seen that quantum mechanics approaches classical mechanics when the energy levels are far from the ground level. In Fig. 8.3 the shape of the probability as a function of the scaled position looks nothing like the classical function, so quantum mechanics gives 'something new'.

8.6.3.1 Application to Statistical Mechanics

This result is more dramatically noted if the result is applied to statistical mechanics. The classical (c) and quantum (q) mechanical partition functions and internal energies are ($\beta = 1/(kT)$)

$$\tilde{Z}_c = \int_{-\infty}^{\infty} dp \int_{-\infty}^{\infty} dx \exp\left[-\beta\left(\frac{p^2}{2m} + \frac{1}{2}m\omega^2 x^2\right)\right] = \frac{2\pi}{\beta\omega}; \quad \tilde{U}_c = \frac{1}{\beta}$$

$$\tilde{Z}_q = \sum_{n=0}^{\infty} \exp\left[-\beta\hbar\omega\left(n + \frac{1}{2}\right)\right] = \frac{e^{-\beta\hbar\omega/2}}{1 - e^{-\beta\hbar\omega}}; \quad \tilde{U}_q = \frac{\omega\hbar}{2}\frac{e^{\beta\hbar\omega} + 1}{e^{\beta\hbar\omega} - 1} \quad (8.78)$$

For high temperature (small β) the quantum mechanical expression for the internal energy behaves as $1/\beta\left[1 + \hbar^2\omega^2\beta^2/12 + O\left(\hbar^4\omega^4\beta^4\right)\right]$ and therefore the quantum mechanical result converges to the classical result for temperatures $T \gg \hbar\omega/k$ when the majority of the occupied levels of the ensemble of oscillators is well away from the ground level.

The above calculation was used by Einstein to estimate the internal energy of a crystal at low temperature. The oscillators are identified as the atoms of the crystal. Many refinements to the calculation can be carried through. First of all, the estimate has to be done in three dimensions; secondly, a correction has to be introduced for the coupling between the oscillators (this was done by Debye, see Mandl (1988)); thirdly, the oscillators need not be isotropic, because the crystal structure may be such that there are directional effects. Taking all these into account, a fairly accurate result for the internal energy—and thereby the specific heat $\tilde{C}_V$—may be obtained.

8.7 Dirac's Ideas

No introduction to quantum mechanics is complete without treating at least some of the ideas about the subject put forward by Paul Dirac (1902–1984)—for a biography see Farmelo (2009). So far, the development of the subject has been focused on finding the wave function $\psi(\mathbf{r})$. It was demonstrated that the solution of Schrödinger's equation leads to quantised bound states, characterised by wave functions $\psi_n(\mathbf{r})$ and energy values E_n; the probability of finding a particle in a space d^3r is $\psi(\mathbf{r})\psi^*(\mathbf{r})$. Physical quantities such as (angular) momentum and energy are obtained by operating on the states. For example, the momentum operator is $\hat{p}_x = (\hbar/i)\partial/\partial x$. The momentum *operator* needs to be distinguished from the momentum *expectation value*, $\langle p_x \rangle$, and also the classical value, p_x, which is why a hat was introduced to denote the operator. The expectation value is

$$\langle p_x \rangle = \int \psi^*(\mathbf{r}) \left(\frac{\hbar}{i} \frac{\partial}{\partial x} \right) \psi(\mathbf{r}) d^3r \tag{8.79}$$

In classical mechanics the distinction between an operator and a value is not necessary, but in quantum mechanics, it is not only necessary but also very useful.

If there are operators there must be objects that are operated on. These objects are obviously the states of the quantum system. If fact, it is convenient to think of it as a vector space. The unit vectors of the space are the states, denoted by $|\psi\rangle$. There is a vector product in the space that defines how operations take place and how vectors are manipulated. Just like in an 'ordinary' (Cartesian) vector space, the operators are matrices and the vectors are characterised by their components and the inner product defines a length, so in quantum mechanics the states are the unit vectors and the operators are more widely defined than matrices and may also include functional prescriptions (as in the example above of the momentum operator).

8.7.1 Operators and Ket Vectors

The space spanned by the unit vectors (each of which denotes a state) is a so-called Hilbert space. Vectors in this space are denoted by $|\psi\rangle$. They are called *ket vectors—*

a strange moniker, at first sight. The inner product of two vectors is made by the bracket $\langle\psi'|\psi\rangle$; the vector $\langle\psi'|$ is called a *bra vector*. Together 'bra' and 'ket' make a bracket—Dirac's little joke. Now, the bra vectors inhabit a different space from the ket vectors because of the complex conjugation that is necessary to make the inner product, as evidenced in the example provided by Eq. (8.79). It implies that

$$\langle\psi'|\psi\rangle = \langle\psi|\psi'\rangle^* \tag{8.80}$$

The inner product, so defined is useful for the investigation of the properties of operators. The ket vectors that are the unit vectors in the Hilbert space, are normalised, so that two unit vectors satisfy the property

$$\langle\psi_a|\psi_b\rangle = \delta_{ab} \tag{8.81}$$

Here the prime has been dropped as the two spaces are clearly distinguished by either $\langle|$ or $|\rangle$, so that it is clear whether one is dealing with a bra or a ket vector.

Hilbert lived from 1862 to 1943 and he conceived of the function space (the Hilbert space) around 1910. Dirac, as well as other mathematical physicists such as von Neumann, began using Hilbert's ideas in the early 1930s. Since then there has been an explosive mathematical research effort, much of which can be applied, to a greater or lesser extent, to quantum mechanics. Within the scope of this text it is not possible to present all, or even most of it. Below, some key results are presented, which are made plausible—but not rigorously proven—by making reference to examples.

The properties of operators may be described by stating what they do with a unit ket vector, denoted by a Latin subscript, for example: $|\psi_a\rangle$. The *matrix element* of an operator $\hat{X}$ is called $\hat{X}_{ab}$ and defined as $X_{ab} = \langle\psi_a|\hat{H}|\psi_b\rangle$. The eigen values λ of an operator $\hat{X}$ are obtained by evaluating

$$\hat{X}|\psi_\lambda\rangle = \lambda|\psi_\lambda\rangle \tag{8.82}$$

The eigen value λ may in general be complex. $|\psi_\lambda\rangle$ are the eigen vectors.

An operator possesses a *Hermitian adjoint* by taking the complex conjugate and transposing. The Hermitian adjoint is denoted by a dagger superscript: †. So for the matrix elements it holds that

$$\hat{X}_{ab}^\dagger = \hat{X}_{ba}^* \tag{8.83}$$

An operator is said to be *Hermitian* when $\hat{X}^\dagger = \hat{X}$. It follows immediately that Hermitian operators have real eigen values.

The link with physical systems is made. The quantum states of a system are—as mentioned above—associated with a ket vector. The design of the Hilbert space is now arranged in such a way that the states are just the unit vectors of the space. If those unit vectors are called $|\psi_n\rangle$, then an arbitrary superposition of states is created by a linear combination of states: $|\psi\rangle = \sum_n c_n|\psi_n\rangle$, where c_n gives the weight of each basic state in a wave packet (or better, state packet). Obviously, $c_n = \langle\psi_n|\psi\rangle$ and it is inferred that $|\psi\rangle = \sum_n |\psi_n\rangle\langle\psi_n|\psi\rangle$, which makes the 'closure' $\sum_n |\psi_n\rangle\langle\psi_n|$ the equivalent of the identity. In principle c_n may be complex and the probability of measuring the basic state labelled n is $|c_n|^2$. The total probability must be $\sum_n |c_n|^2 = 1$.

There is quite a bit of freedom in choosing the unit vectors. For example, the unit vectors may be given by Schrödinger's equation; they are the eigen vectors of the Hamiltonian operator

$$H|\psi_n\rangle = E_n|\psi_n\rangle \tag{8.84}$$

A word of warning is appropriate here. Ket and bra vectors are *not* Cartesian vectors, even though quantum practitioners tend to the same vocabulary: unit vectors, eigen vectors, etc.; they do not transform in the same way under a physical rotation.

8.7.1.1 The Wave Function and Ket Space

The notation of the ket vectors may be quite loose, for instance, the unit vectors of the one-dimensional harmonic oscillator are frequently denoted by $|n\rangle$. In the treatment of the various examples of Schrödinger's equation in the preceding sections the spatial wave functions have been determined. Now, it was noted in the discussion on the free particle that a spatial description is as valid as a momentum description. The 'quantum state' is obviously something more encompassing than merely the spatial wave function. The quantum states, as characterised by ket vectors, prefer no particular representation and the question must be: how can a spatial wave function be obtained from a ket vector that characterises a state? To that end spatial eigen vectors must be determined, in other words, if the position operator $\hat{x}$ works on a state then the eigen value comes out as the position x. (For simplicity—and just to illustrate the issue—a one-dimensional case of one particle is considered here.) The eigen vectors are the ket vectors $|x\rangle$

$$\hat{x}|x\rangle = x|x\rangle \tag{8.85}$$

The eigen vectors span the Hilbert space and therefore quantum states may be written as weighted sums over the vectors $|x\rangle$. Here is a difficulty: the variable x is continuous, it is not a discrete set of eigen vectors and eigen values. Therefore, instead of a sum, an integral over the eigen states must be taken.

$$|\psi\rangle = \int dx\, c(x)|x\rangle \tag{8.86}$$

where now the discrete set of values of c_n has been replaced with a continuous function $c(x)$. The closure relation also becomes an integral rather than a sum and the discrete identity, the Kronecker delta, becomes a Dirac delta function

$$\langle x|x'\rangle \rightarrow \delta(x - x') \tag{8.87}$$

No equal sign can be used here, as the delta function only makes sense under an integral. The function $c(x')$ is now interpreted as the wave function

$$\langle x|\psi\rangle = \int dx'c(x')\langle x|x'\rangle = \int dx'c(x')\delta(x-x') = c(x) \quad \rightarrow \quad \psi(x) = \langle x|\psi\rangle$$
(8.88)

Entirely analogously the momentum representation of a quantum state that is characterised by the ket vector $|\psi\rangle$ is given by

$$\phi(p) = \langle p|\psi\rangle$$
(8.89)

Physical quantities are obtained by letting operators work in ket space, that is, a physical quantity q is obtained from a ket vector by evaluating $\hat{q}|\psi_n\rangle = q|\psi_n\rangle$. This is the definition of an eigen vector and as these must be real, the operator $\hat{q}$ is Hermitian. Thus, *operators that correspond to physical quantities are Hermitian*.

8.7.2   Commutators

The result of a multiplication of two operators depends on the order in which the multiplication is done. By creating the difference between the product of $\hat{X}\hat{Y}$ and $\hat{Y}\hat{X}$ it is possible to evaluate the extent of this feature. The difference $\hat{X}\hat{Y} - \hat{Y}\hat{X}$ is denoted by $\left[\hat{X}, \hat{Y}\right]$ and is called the *commutator of* $\hat{X}$ and $\hat{Y}$. It is almost impossible to do any serious quantum mechanics without considering commutator rules.

As an example, consider the time evolution of a physical parameter. Let this parameter be given by the operator $\hat{X}$, then its expectation value is

$$\langle\hat{X}\rangle(t) = \int \psi^*(x,t)\hat{X}(t)\psi(x,t)dx$$
(8.90)

The time derivative is

$$\frac{d}{dt}\langle\hat{X}\rangle = \int \psi^*(x,t)\frac{\partial\hat{X}}{\partial t}\psi(x,t)dx +$$
$$+ \int \frac{\partial\psi^*(x,t)}{\partial t}\hat{X}\psi(x,t)dx + \int \psi^*(x,t)\hat{X}\frac{\partial\psi(x,t)}{\partial t}dx$$
(8.91)

Using Schrödinger's equation, as well as its complex conjugate

$$\frac{d}{dt}\langle\hat{X}\rangle = \frac{\partial}{\partial t}\langle\hat{X}\rangle + \frac{i}{\hbar}\int \psi^*(x,t)\hat{H}\hat{X}\psi(x,t)dt - \frac{i}{\hbar}\int \psi^*(x,t)\hat{X}\hat{H}\psi(x,t)dt =$$
$$= \frac{\partial}{\partial t}\langle\hat{X}\rangle + \frac{i}{\hbar}\langle\left[\hat{H}, \hat{X}\right]\rangle$$
(8.92)

So, it follows that the commutator of a physical parameter operator with the Hamiltonian describes the time evolution of that parameter. Reverting back to classical mechanics, it was shown in Sect. 5.3.2 that the time evolution of a parameter is connected to the *Poisson bracket*, see Eq. (5.31)

$$\frac{df}{dt} = \frac{\partial f}{\partial t} + \{f, H\} \qquad (8.93)$$

So, the Poisson bracket in classical mechanics and the commutator in quantum mechanics play similar roles. The implication is that a quantity that does not explicitly depend on time, is a *conserved quantity if it commutes with the Hamiltonian*. Dirac has shown that the classical equivalent of *any* commutator (and not just the Hamiltonian) is the Poisson bracket.

Applying Eq. (8.92) to the momentum operator $(\hbar/i)\partial/\partial x$ one finds

$$\int \psi^*(x)\frac{dp}{dt}\psi(x)dx = \frac{i}{\hbar}\int \psi^*(x)\left[-\frac{\hbar^2}{2m}\frac{\partial^2}{\partial x^2} + V(x), \frac{\hbar}{i}\frac{\partial}{\partial x}\right]\psi(x)dx \quad (8.94)$$

The integrand in the right-hand member is developed as

$$\psi^*(x)\left[-\frac{\hbar^2}{2m}\frac{\partial^2}{\partial x^2} + V(x), \frac{\hbar}{i}\frac{\partial}{\partial x}\right]\psi(x) = -\frac{\hbar}{i}\psi^*(x)\frac{\partial V(x)}{\partial x}\psi(x) \qquad (8.95)$$

Therefore,

$$\frac{d\langle p \rangle}{dt} = -\langle\frac{\partial V(x)}{\partial x}\rangle \qquad (8.96)$$

This finding, with which the name of Ehrenfest is associated, points to Newton's second law and is an example of the classical limit, or as it is called in quantum mechanical jargon the *correspondence principle*. The result is here demonstrated in one dimension, but it is easy to show that it works just as well in three dimensions.

The commutator plays another important role in quantum mechanics. An example of two non-commuting operators is the position and the momentum $[\hat{x}, \hat{p}] \rightarrow \hat{x}\hbar/i\,(\partial/\partial x) - \hbar/i\,(\partial/\partial x)\hat{x} = -\hbar/i$ (just let it work in position representation in which $\hat{x} \rightarrow x$ and $\hat{p}_x \rightarrow (\hbar/i)\partial/\partial x$). Now, Heisenberg's uncertainty relation states that momentum and position cannot be measured simultaneously. And this points to a general result: *two quantities can be measured simultaneously when their operators commute*. More mathematically, it may be shown that if two operators commute, they can be diagonalised simultaneously, in other words they share a set of eigen vectors.

In passing it is noted that for the angular momentum operators $\hat{L}_i = \epsilon_{ijk}\hat{x}_j\hat{p}_k$ it holds that

$$[\hat{L}_x, \hat{L}_y] = i\hbar\hat{L}_z \,, \quad [\hat{L}_y, \hat{L}_z] = i\hbar\hat{L}_x \,, \quad [\hat{L}_z, \hat{L}_x] = i\hbar\hat{L}_y \qquad (8.97)$$

Dirac made many other discoveries in quantum mechanics, very especially he combined quantum mechanics with relativity theory to describe spin and in so doing—quite extraordinarily—predicted that each particle should have an anti (or mirror) particle. These anti particles have an opposite charge to the particle, for example, the positron (anti-electron) has exactly the same, but opposite charge to the electron (and the same mass). Some particles possess their own anti particle. Such particles necessarily have a zero charge. The photon is an example of this.

8.8 Ladder Operators

The Hamiltonian operator for the one-dimensional harmonic oscillator is

$$\hat{H} = \frac{\hat{p}^2}{2m} + \frac{1}{2}m\omega^2\hat{x}^2 \tag{8.98}$$

Introducing the following operators, leads to interesting insights

$$\hat{a} = \sqrt{\frac{m\omega}{2\hbar}}\left(\hat{x} + i\frac{\hat{p}}{m\omega}\right); \ \hat{a}^\dagger = \sqrt{\frac{m\omega}{2\hbar}}\left(\hat{x} - i\frac{\hat{p}}{m\omega}\right) \tag{8.99}$$

Now, using $\left[\hat{x}, \hat{p}\right] = i\hbar$, it follows easily that

$$\hat{H} = \hbar\omega\left(\hat{a}\hat{a}^\dagger - \frac{1}{2}\right) = \hbar\omega\left(\hat{a}^\dagger\hat{a} + \frac{1}{2}\right) \ and \ \left[\hat{a}^\dagger, \hat{a}\right] = -1 \tag{8.100}$$

The eigen vectors of $\hat{H}$ are obviously the same as the ones of $\hat{a}^\dagger\hat{a}$; these are denoted by $|n\rangle$ and the eigen values are called λ_n:

$$\hat{a}^\dagger\hat{a}|n\rangle = \lambda_n|n\rangle \tag{8.101}$$

The values of λ_n must be positive or zero as follows from the fact that any norm of a vector is positive

$$\lambda_n\langle n|n\rangle = \lambda_n = \langle n|\hat{a}^\dagger\hat{a}|n\rangle = \langle \hat{a}n|\hat{a}n\rangle \geq 0 \tag{8.102}$$

Now construct

$$\left(\hat{a}^\dagger\hat{a}\right)\hat{a}^\dagger|n\rangle = \hat{a}^\dagger\left(\hat{a}^\dagger\hat{a} + 1\,\middle|\,n\right) = (\lambda_n + 1)\hat{a}^\dagger|n\rangle \tag{8.103}$$

In other words, not only is $\hat{a}^\dagger|n\rangle$ another eigen vector of $\hat{a}^\dagger a$, its eigen value is the eigen value belonging to $|n\rangle$ plus 1. In the same way, it is shown that the operator $\hat{a}$ reduces the eigen value by 1. So, by letting $a^\dagger$ or a work—possibly repeatedly—other eigen vectors are found with either increasing or decreasing eigen values. Therefore,

$a^\dagger$ is called the *creation* or *raising* operator and a the *annihilation* or *lowering* operator. Such operators are known as *ladder operators*, for obvious reasons.

As the smallest value of λ_n equals 0, applying the lowering operator time and again would reduce the eigen value until it reaches a lowest value. This value is necessarily equal to zero and the eigen vector that corresponds to it is called $|0\rangle$. By applying the creation operator any number of times the eigen vectors $|n\rangle$ are obtained. The successive eigen values are 0, 1, 2, 3 .... n. The energy levels are then, keeping Eq. (8.100) in mind

$$E_n = \hbar\omega\left(n + \frac{1}{2}\right) \tag{8.104}$$

This result has been achieved without the need to solve a differential equation.

All the eigen vectors can be obtained from the lowest 'null' or vacuum state $|0\rangle$ by repeatedly applying the creation operator. The states $|n\rangle$ have to be normalised—$\langle n|n\rangle = 1$—and it is left as an exercise for the reader to show that

$$|n\rangle = \frac{a^{\dagger n}}{\sqrt{n!}}|0\rangle \tag{8.105}$$

Furthermore,

$$a^\dagger|n\rangle = \sqrt{n+1}|n+1\rangle \text{ and } a|n\rangle = \sqrt{n}|n-1\rangle \tag{8.106}$$

At the beginning of this chapter the identification of a particle with a wave was made. Going a step further, the identification of a *state* with a particle is made. Thus, the excited states of the harmonic oscillator are 'particles', called *phonons* or *excitons*. Vibrations in the electromagnetic field, *photons*, can be similarly described using creation and annihilation ladder operators. In fact, all elementary particles may be described as such and in this way Dirac laid the foundation for the *field theory* that deals with processes, such as particle collisions as experimented with in high energy colliders, in which various species of particles are exchanged.

8.9 Measurement and Quantum Mechanics

The measurement of quantities of a physical system involves the interaction of the system with something that can be sensed and quantified by a human being. Any measurement brings with it some influence of the system that is being measured. In classical mechanics that is not such a problem. Planets can be seen in the sky as light from the Sun illuminates them, some of which bounces off to activate a human eye, who then 'sees' the planet as a light dot. The light from the Sun is not considered to influence the orbit of the planet unduly and therefore the observation of the planet is 'true'. For smaller systems, for example, the measurement of a milliamp current in an electrical circuit, the question of influence of the measurement apparatus on the value of the current *is* more relevant. A tiny fraction of the current has to be diverted from the circuit into the amp meter to affect a motion of the needle. A careful calculation has to be done to relate the 'truth' of the reading to the actual value of the current in the circuit. If the meter uses too much of the current, to such an extent that the functioning of the circuit is substantially affected, then the measurement does not make any sense.

While the above sentences may seem overly philosophical, it is a fact that a quality undergraduate course in physics includes a large amount of practical work, which is concerned not only with demonstrating 'effects' but also with extended estimates of errors of measurement and assessments of the accuracy of measured values. It is part and parcel of the subject of physics, essential to its intellectual rigour, even though it is not very glamorous and hardly ever even mentioned in popular expositions.

The measurement of a quantum system presents additional problems. Somehow the quantum process has to move a needle, blacken a photographic plate or leave a trace to betray its presence. Needles, photographic plates and traces are all macroscopic objects, ruled by classical mechanics. To obtain a measurement the quantum system, therefore, needs to interact with a classical system and do so preferably without its properties being unduly influenced. Classical systems have many particles and are affected by outside influences, such as temperature fluctuations. The latter are generally so large (in terms of energy content) that any quantum system is bound to be disrupted by a substantial amount. If light is used to detect quantum particles (electrons, for example), then the accuracy to which their position can be ascertained depends on the wavelength of the radiation that is applied. For higher accuracies shorter wavelengths are required, which increases the energy of the photons and thus affects the quantum system to a greater extent. This may be quite pernicious.

As an example consider the famous *two slit experiment*. A beam of either classical projectiles—bullets, for example—or quantum objects—monochromatic light or electrons—is fired at a screen with two slits in it, see Fig. 8.4. Behind the screen is an array of detectors, that are designed as appropriate to the particles that are emitted from the source. Electrons have a wave character and therefore an interference pattern is detected on a photographic plate (that is, the relevant detector), which is positioned behind the screen. For the classical objects the shadows of the two slits are revealed, so the detectors 'see' two bumps. For the quantum objects an interference pattern is found. Now, electrons *also* have a particle character and therefore the question as to which electron goes through which slit is important if one wants to understand how the interference pattern has come into being. If light is used to detect the path of the electrons, then it transpires that if the wavelength of the light is sufficiently small to be able to distinguish the distance between the two slits, the influence of the photons on the electron beam is so great that the interference pattern disappears!

The question is now: how can a classical system detect quantum particles without undue influence on the quantum process? Ingeniously, mankind has found a way to do this. A classical system is set up in a *metastable state*. For example, in a *bubble chamber* a fluid is brought in a state of supersaturated vapour, see Sect. 7.7.3. The classical system, which consists of many particles, is normally in a state of thermodynamic equilibrium, but in a metastable state, which can exist for long periods of time, a miniscule fluctuation can bring about a macroscopic transition,

Fig. 8.4 Illustration of the two-slit experiment. S is the source from which either classical objects (bullets, say), monochromatic photons or electrons are emitted. T is the screen with two slits in it and D is the detector. The patterns that is detected for the classical objects (c) is two bumps, but for light or electrons (q-l) it is an interference pattern

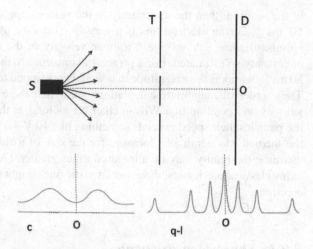

which can then be detected and recorded. The bubble chamber (also known as a Wilson chamber) is such a system and the motion of quantum particles is seen as a trace of bubbles. Similarly, in a Geiger counter a dilute perfect gas is brought in a metastable conductive state in a *Geiger-Müller tube*, which, as a consequence of a tiny perturbation caused by a passing particle, turns the gas into an electrical conductor for a very short period of time, which is then recorded via electronic amplification as a 'click'. Photographic plates contain metastable AgBr crystals, which, again, are excited by a passing small fluctuation. Basically, metastable systems, once triggered, are irreversible. In this way, quantum systems may interact with classical systems.

Taking the example of the Wilson chamber, in which the tracks of particles are made visible, questions arise. A high-speed film may be made of the bubble path, revealing where the particle has been and how fast it has travelled. In passing it is noted that a magnetic field may be applied to reveal the electrical properties of the particles by producing curved orbits. Is not the fact that position and velocity are determined at the same time in contradiction with Heisenberg's fundamental uncertainty relation? To solve that puzzle an assessment has to be made of the accuracy of the position measurement in the bubble chamber. It would not be unreasonable to assume that the accuracy is no greater than the *thermal wavelength* of the fluid in the chamber. After all, for bubbles to form, a collective phase transition has to take place, which involves a quantity of fluid of many particles. For a van der Waals gas it was derived (see Sect. 7.7.3) that the thermal wavelength is of the order of

$$\lambda_T = \left(\frac{\beta h^2}{2\pi m}\right)^{1/2} \tag{8.107}$$

Some indicative numbers are used: the temperature is some $300\,\mathrm{K}$ and the particle mass of the fluid is of order of a water molecule $m \approx 3 \times 10^{-26}\,\mathrm{kg}$, which leaves an approximate value of λ_T of $10^{-11}\,\mathrm{m}$. If ten times this value is used as the uncertainty in the position, then the uncertainty in the velocity for an electron comes to some $10^6\,\text{ms}^{-1}$. For an electron this is a very low velocity, as typically it would take a voltage of some 3 V to give it a mean velocity of the order of magnitude of the uncertainty so estimated. For a proton the uncertainty in velocity is of the order of $50\,\text{ms}^{-1}$, which is the magnitude that would correspond to a voltage of some 0.15 V. These order of magnitude calculations show that there is no problem measuring velocity and position in a Wilson chamber as long as the applied voltages to give the particles their speed exceeds something like 10 V—a very practical value. Using the thermal wavelenth as a measure for the size of a bubble is also a conservative estimate; the reality may be a hundred times greater. (All estimates are calculated using classical mechanics, though relativistic ones might be more appropriate for the electron).

8.9.1 Sharp Measurements

Quantum mechanics makes probabilistic predictions and this leads to issues, viewpoints and discourses, which have attracted a great deal of philosophical debate, not only about the nature of the subject but also about the reality of the world experience. One manner of interpreting the statistical aspect of the subject is to do a *series of experiments*. In this scenario, a certain quantum state is prepared, followed by the measurement of a certain variable, for example, the position of a particle. Then the same state is prepared again and another measurement of the same variable is taken. This process is repeated numerous times and a statistic can be made of the values of the variable. An average and a standard deviation (or variance) are calculated, which may be compared to the theoretical predictions (by solving the Schrödinger equation appropriate to the prepared state). It would seem that quantum mechanics pursued in this way is entirely consonant with the foundational intentions of the subject.

In the context of the same scenario, an analysis may be carried out. Suppose the quantum state is prepared so that the wave function is $\psi(x, t)$ (or more compactly, the ket $|\psi\rangle$) and the variable to be measured is represented by an operator $\hat{p}$. The expectation value of the variable at time t is

$$\langle P \rangle = \int \psi^*(x, t)\hat{p}\psi(x, t)dx \qquad (8.108)$$

The variance is

$$\langle \Delta P \rangle = \int \psi^*(x, t)\left(\hat{p} - \langle P \rangle\right)^2 \psi(x, t)dx \qquad (8.109)$$

this may be further developed as

$$\langle \Delta P \rangle = \int \left[\left(\hat{p} - \langle P \rangle\right)\right]^* \left(\hat{p} - \langle P \rangle\right) \psi(x, t)dx \qquad (8.110)$$

which vanishes when $\hat{p}|\psi\rangle = \langle P\rangle|\psi\rangle$. In that case the variance is zero and the value given by the operator $\hat{p}$ is said to be *sharp*.

These findings have profound implications. If a measurement is done on a quantum state a certain value is found. If the measurement is then repeated immediately the reasonable physical assumption must be that the same value is obtained. This means that after a measurement the state is sharp and therefore *measurement forces a quantum state into an eigen state*. More precisely, suppose that a quantum state $|\psi\rangle$ is a superposition of eigen states $|\psi_i\rangle$, that is

$$|\psi\rangle = \sum_i c_i|\psi_i\rangle \tag{8.111}$$

then on measurement the probability of finding the state in $|\psi_i\rangle$ is c_i^2. The measurement will give some value, say P_j, and then suddenly the state collapses into the eigen state that belongs to the eigen value of the operator $\hat{p}$ that is associated with the quantity that is measured and forces the system into the state $|\psi_j\rangle$ for which $\hat{p}|\psi_j\rangle = P_j|\psi_j\rangle$.

8.9.2 Schrödinger's Cat

Schrödinger sought to explain this aspect of quantum mechanics in a humorous way by means of an illustration on a macroscopic object: the (in)famous *Schrödinger's cat*.

In Schrödinger's thought experiment a cat is locked in a box. Also in the box is a radioactive sample of material, a vial of poison and a mechanism with a hammer. When the radioactive sample transits from one state to another, a mechanism releases the hammer, which smashes the vial of poison and the cat dies. Note that all elements of a quantum measurement are represented here. The actual quantum mechanical process is the radioactive material. The mechanism with the hammer is in the metastable state, which is irreversible because once the vial is smashed it cannot go back to its initial state; it cannot be un-smashed. A small peep hole allows an observer to check whether the cat is alive or dead. In a primitive way, then, the system has two eigen states: the cat can be alive or dead. While no measurement has been made the cat is simultaneously in two states, that could be represented as $|\text{cat}\rangle = c_1|\text{alive}\rangle + c_2|\text{dead}\rangle$. Looking through the peep hole forces the system in either of the eigen states. It could be said that the act of observation determines the state that the cat is in.

The problem comes when the observation process is not straightforward. The observer may delegate the act of looking through the peep hole to a family member or work colleague, who then later on informs the observer of the result. There could be a whole chain of intermediaries, so the question is: when does the observation take place and who is the observer? Is that the person who first looks, or the one who finally publishes the result? More esoterically even, the decision to look determines the state of the cat; it is as if the state of mind of the observer has decided on the cat's fate! This would seem to be the status of the subject. That is strange, because suddenly, quantum mechanics has become the subject of the state of mind of the observer.

8.9.3   Van Kampen's Apparatus

Various notable physicists have opposed this interpretation of quantum mechanics. Especially Einstein was not at all in favour. Others have pointed out that the measurement apparatus should be part of the description. One vociferous proponent of such a course of action was van Kampen. In a paper entitled *Ten theorems about quantum mechanical measurements* (Van Kampen 1988) he analyses the so-called *Copenhagen interpretation of quantum mechanics* and introduces a method for including the measurement apparatus into the story of measurement.

Van Kampen's apparatus is envisaged as follows. An atom is brought into an excited state. In order to transit into the ground state it must emit a photon. The photon has angular momentum and, as the angular momentum is a conserved quantity, the atom's angular momentum difference between the excited state and the ground state must be such that it exactly matches the angular momentum of the photon. In the apparatus the excited state is chosen in such a way that the balance is not satisfied and therefore the excited state cannot of itself transit to the ground state. Now, when an electron from a quantum process passes by, a dipole moment is introduced, which makes the transition to the ground state possible and the atom can emit its photon. In this way a metastable state is created. The photon may be detected by a photographic plate, which is also part of the apparatus. Van Kampen describes the state of the apparatus in terms of a Schrödinger equation, making various plausible assumptions. Then it is possible to write down the state before the electron interacts with the excited atom, as well as the one after the interaction has taken place (and the photon has presumably been detected by the photographic plate).

While this procedure would appear to describe the collapse of the wave function into an eigen state, it introduces a new uncertainty, because after the interaction the apparatus will be in a new—and different—superposition. The latter needs to be determined in another measurement. The measurement problem has thus been shifted from a simple one—that is, merely the electron flying by with uncertainty in its position—to a much more complex one, the new state of the (densely spaced) energy levels of the system electron+apparatus.

Another question would be: where does the apparatus end? The experimental observation may be carried out by means of a magnifying glass to read the blackening of the photographic plate. Now the magnifying glass is part of the system, which would require an extension of the Schrödinger equation to include it. And so it could go on and finally the whole universe may need to be included. There is no real solution to this problem; it is an integral part of quantum mechanics. It could be argued that such a statement is fatalistic and even contrary to the purposes of science. However, the statement does not imply that it is forbidden to think about it and unfortunately some have argued for that.

8.9.4 Systems of More Than One Particle

The measurement problem becomes distinctly interesting when systems of more than one particle are studied.

8.9.4.1 Two Independent Particles in a Box

As a simple example first consider two particles in a box. In one dimension, and assuming that the particles can be distinguished, the subscript 1 is attached to operators relevant to particle number 1 and 2 to the other particle. The Schrödinger equation reads

$$\left(-\frac{\hbar^2}{2m_1}\frac{\partial^2}{\partial x_1^2} - \frac{\hbar^2}{2m_2}\frac{\partial^2}{\partial x_1^2}\right)\psi(x_1, x_2) = E\psi(x_1, x_2) \tag{8.112}$$

The solution is obtained by setting $\psi(x_1, x_2) = \psi_1(x_1)\psi_2(x_2)$ and $E = E_1 + E_2$. The equation then splits into two independent equations

$$-\frac{\hbar^2}{2m_1}\frac{\partial^2\psi(x_1)}{\partial x_1^2} = E_1\psi(x_1) \tag{8.113}$$

$$-\frac{\hbar^2}{2m_2}\frac{\partial^2\psi(x_2)}{\partial x_1^2} = E_2\psi(x_2) \tag{8.114}$$

Boundary conditions—for example, the ones relevant to one particle in a box—can be applied independently as well. So, there would seem to be no problem. If the location of one of the particles is measured, the location of the other particle may be measured independently. The essential property that ensures the independence of the particles here is that $\psi(x_1, x_2) = \psi_1(x_1)\psi_2(x_2)$, or, in ket-language $|\psi(x_1, x_2)\rangle = |\psi(x_1)\rangle|\psi(x_2)\rangle$.

8.9.4.2 Two Interacting Particles in a Box

But what would happen if that is not the case, if no direct product describes the state of the particles. This scenario would apply if the particles are somehow interacting, or have interacted in the past. One possible way of examining such a scenario is one in which the centre of gravity motion and relative motion are considered. Calling

176   8 Elements of Quantum Mechanics

the positions of the two particles x_1 and x_2, as before and the relative coordinate is $r = x_1 - x_2$ and the centre of gravity coordinate is $R = (x_1 + x_2)/2$. The analysis here is a first-order illustration; many refinements are possible. Let the wave function of the combined system take the form $\psi(R, r) = \phi(R)\rho(r)$; furthermore assume that the centre of gravity part of the wave function is symmetric: $\phi(-R) = \phi(R)$ (while this may appear rather restrictive, it still covers a wide range of physically realistic cases). The position of particle 1 may be measured and let the outcome of that measurement be $x_1 = a$. Now, it was found before that a measurement forces the system into an eigenstate, in other words the operator $\hat{x}_1$ has an eigen value a, associated with an eigen vector ψ_a, or $\hat{x}_1|\psi_a\rangle = a|\psi_a\rangle$. Then

$$\int \psi^*(\tfrac{1}{2}(x_1 + x_2), x_1 - x_2)\hat{x}_1\psi(\tfrac{1}{2}(x_1 + x_2), x_1 - x_2)dx_1dx_2 =$$

$$= a \int \psi^*(\tfrac{1}{2}(x_1 + x_2), x_1 - x_2)\psi_a(\tfrac{1}{2}(x_1 + x_2), x_1 - x_2)dx_1dx_2 \quad (8.115)$$

The right-hand side of this equation equals a, as the eigen vectors have a magnitude of unity and they are orthogonal, so the operator $\hat{x}_1$ 'selects' the eigen vector $|\psi_a\rangle$. Of course, the operator $\hat{x}_1$ is known; it is the coordinate itself x_1.

What would be the value measured for x_2? In other words, what is the outcome of the expression

$$\int \psi^*(\tfrac{1}{2}(x_1 + x_2), x_1 - x_2)\hat{x}_2\psi(\tfrac{1}{2}(x_1 + x_2), x_1 - x_2)dx_1dx_2 \quad (8.116)$$

Again, the operator $\hat{x}_2$ is known; it is the coordinate x_2. In the integral the transformation $x_1 = -x_2$ and $x_2 = -x_1$ is made. The product dx_1dx_2 remains invariant, as does the relative coordinate $r = x_1 - x_2$. Furthermore, as it was assumed that $\phi(-x1 - x_2) = \phi(x_1 + x_2)$, the integral takes the form

$$- \int \psi^*(\tfrac{1}{2}(x_1 + x_2), x_1 - x_2)x_1\psi(\tfrac{1}{2}(x_1 + x_2), x_1 - x_2)dx_1dx_2 \quad (8.117)$$

The outcome is precisely the same as that for the operator $\hat{x}_1$, other than a minus sign. So, the eigen vector for this measurement is identical, but the eigen value has changed sign.

This result is significant. What it shows is that a measurement on the position of one of the particles, immediately determines the position of the other particle, even if the particles are far apart. That would suggest some form of 'superluminal' communication between the particles, which would appear to fall foul of relativity theory. In principle it could be argued that the form of the wave function suggested here is erroneous, as it would be the solution of a Schrödinger equation with an instantaneous interaction, whereas a time-retarded interaction should have been used. However, the instantaneous effect is also evident in situations in which particles carry intrinsic properties, such as particle spin. If the spin of one particle is measured (and if the experiment is prepared in such a way that the total spin—intrinsic angular momentum—is zero), then instantaneously the spin of the other particle is no longer arbitrary. Einstein, in particular, was very worried about this aspect of quantum mechanics and a long—and some might say, muscular—discussion followed between the 'quantum heavy-weights' Einstein, Schrödinger and Bohr. Einstein referred to *entanglement* and its, admittedly, disconcerting as 'spooky action at a distance'. His idea was that quantum mechanics is somehow incomplete; there must be 'hidden variables' that define the state better than the wave function can. In a famous paper with co-authors Podolsky and Rosen he described his misgivings, Einstein et al. (1935).

While the example above is somewhat contrived, it is instructive. If the position of the particles is determined with great precision then the momentum must—by the uncertainty relation—be indeterminate, which implies that the range of momenta must be large; this in turn makes the energy content of the system very large. If the particle pair did not have this energy before the measurement, then it follows that it has been transferred from the measurement apparatus. This is the case when the measured property is the position. If the measured property is intrinsic to the particle however, the transfer of energy to the system need not be significant and therefore it is worthwhile considering how exactly an intrinsic property becomes manifest in a quantum context.

8.9.5 Particle Spin

As an example take the property of spin. This feature of a quantum particle is often discussed in the popular literature as if a particle is a spinning top, but the way quantum spin becomes manifest has features that defy the classical image of a spinning top. If it were something like a spinning top then there would be an axis around which the spin takes place, which gives the particle angular momentum. It is certainly true that properties akin to angular momentum are associated with spin; for example, its magnitude can be characterised in terms of $\hbar$, which has the dimension of angular momentum. For electrons the spin property couples to a magnetic field and so it may be measured. Naively, one would expect that all values between a maximum and minimum may be attained, depending on the angle between the magnetic field and the direction of the spin vector of the particle at the time of measurement. But no, for an electron only two values are ever found: $\hbar/2$ and $-\hbar/2$. Also, keeping in mind that the axis of the intrinsic property may have any orientation in space, it would not be unreasonable to expect that three components may be measured, but this is also not true. No experiment can be devised in which all three components may be determined at the same time. This is rather similar to the situation in which position and momentum of a particle are measured; the uncertainty relation states that these two variables cannot be determined simultaneously. Spin, therefore, is a typical 'quantum property' and crucially *the role of measurement is essential in understanding it*. Measurements on the spins of a particle pair have resulted in a much better understanding of quantum mechanics. These measurements were first done by Stern and Gerlach and these are a rather exciting chapter in the history of physics. A beam of particles is created and guided through a magnetic field, which then splits into two, due to the coupling to the spin and this experiment reveals spin's strange quantum property.

The question is now how such properties are described mathematically. As the spin property is intrinsic and not coupled to the position or momentum of a particle, the wave function can just be enhanced to account for it by multiplication. So, instead of $A \exp(ipx/\hbar)$, introduce something like (spin property) $\exp(ipx/\hbar)$. From the fact that one component of the spin angular momentum is always equal to either $\hbar/2$ or $-\hbar/2$ it may be inferred that the spin property in front of the e-power has the form of a vector and that this vector is an eigen vector, while $\pm\hbar/2$ are the eigen values of an operator. This operator is necessarily a matrix. Now, it must be noted that the vector in question is *not* a vector in a spatial sense, but rather a vector in ket space. So, setting the vectors that belong to the two possible measurement outcomes in a chosen direction (say the z-direction) as [1, 0] and [0, 1], then the matrix is diagonal such that

$$\frac{\hbar}{2}\begin{bmatrix}1\\0\end{bmatrix} = \frac{\hbar}{2}\begin{bmatrix}1&0\\0&-1\end{bmatrix}\begin{bmatrix}1\\0\end{bmatrix} , \quad -\frac{\hbar}{2}\begin{bmatrix}0\\1\end{bmatrix} = \frac{\hbar}{2}\begin{bmatrix}1&0\\0&-1\end{bmatrix}\begin{bmatrix}0\\1\end{bmatrix} \tag{8.118}$$

The state belonging to [1, 0] is generally referred to as 'spin up', while the one that belongs to [0, 1] is 'spin down'. Again, these are not spatial indications, but the vectors represent states; the spin up state could be denoted by $|u\rangle$ and the spin down state as $|d\rangle$. A mixed state would have the form $c_u|u\rangle + c_d|d\rangle$, with the probability of finding the up or down states equal to $|c_u|^2$ and $|c_d|^2$, respectively. The matrix operator that produces the experimentally obtained eigen values for a measurement of the z-component of the spin is called $\hat{S}_z$. The other components of the spin are associated with matrices $\hat{S}_x$ and $\hat{S}_y$. As it is an angular momentum operator the commutation rules (8.97) apply and take the form

$$[\hat{S}_x, \hat{S}_y] = i\hbar\hat{S}_z , \quad [\hat{S}_y, \hat{S}_z] = i\hbar\hat{S}_x , \quad [\hat{S}_z, \hat{S}_x] = i\hbar\hat{S}_y \tag{8.119}$$

These relations are satisfied for the following forms of the operators

$$\hat{S}_x = \frac{\hbar}{2}\begin{bmatrix}0&1\\1&0\end{bmatrix} , \quad \hat{S}_y = \frac{\hbar}{2}\begin{bmatrix}0&-i\\i&0\end{bmatrix} , \quad \hat{S}_z = \frac{\hbar}{2}\begin{bmatrix}1&0\\0&-1\end{bmatrix} \tag{8.120}$$

The eigen vectors that belong to $\hat{S}_x$ and $\hat{S}_y$ are

$$\hat{S}_x : \quad \frac{1}{\sqrt{2}}\begin{bmatrix}1\\1\end{bmatrix}, \begin{bmatrix}-1\\1\end{bmatrix} \quad \hat{S}_y : \quad \frac{1}{\sqrt{2}}\begin{bmatrix}-i\\1\end{bmatrix}, \begin{bmatrix}i\\1\end{bmatrix} \tag{8.121}$$

It is now of interest to see what happens when two Stern-Gerlach experiments are done in sequence. First an experiment is done that acts on the z-direction. A beam of electrons is split into two, one has spin up and the other spin down. Then a second Stern Gerlach experiment is set up that captures one of the beams, for example, the spin up beam, but it splits the beam in the x-direction. So, on entering the second apparatus the spin vector is an eigen vector of $\hat{S}_z$ and has the form $[1, 0]$. What does $\hat{S}_x$ do with this vector? It splits the beam into two with eigen vectors $[1, 1]/\sqrt{2}$ and $[-1, 1]/\sqrt{2}$, so all information that the $\hat{S}_z$ has put into the beam is eradicated. It is easy to see that the expectation value of a spinor $[1, 0]$ from the operator $\hat{S}_x$ is zero, but the standard deviation does not vanish and therefore $[1, 0]$ is not an eigen vector of $\hat{S}_x$.

In passing it is noted that the eigen vectors of $\hat{S}_x$ make an agle of $\pi/4$ with those of $\hat{S}_z$. This is an angle in ket space. The spatial angle between the z state and the x state is twice that: $\pi/2$. Performing the rotation four times in physical space will only have rotated the ket space angle by half that, essentially turning spin up into spin down and vice versa. This is a very strange result from a classical point of view, as it would normally be expected that a rotation of 2π would return the system to its original state. It is an unexpected example of the way in which quantum mechanics deviates from classical mechanics. For the spin to return to its original state two full rotations have to be carried out. The fact that two spin states can be obtained from one another by a rotation in ket space will be used below.

The above analysis, which is valid for electrons and protons, pertains to particles that possess the property of a spin value that is either $\hbar/2$ or $-\hbar/2$. Such particles are said to have 'spin 1/2'. There are other particles that also have spin, but not necessarily spin 1/2. Generally, particles have a spin value of $n\hbar/2$, where $n = 0, 1, 2, 3, \ldots$. If n is odd, such as in the case of the electron (where $n = 1$), the particles are fermions and obey Fermi-Dirac statistics, ruled by Pauli's exclusion principle that states that each energy level in the system can only be filled with one set of quantum numbers (the spin itself is also a quantum number, so for example, in the one-dimensional box or the harmonic oscillator each energy level can be occupied by two electrons: spin $\hbar/2$ and spin $-\hbar/2$). For a particle that has spin $s\hbar$, there are $2s + 1$ possible spin values; in the case of the electron $s = 1/2$, corresponding to two spin values. A spin 1 particle has three possible states, etc. When n is even (including $n = 0$) the particles are bosons and obey Bose-Einstein statistics, which lets each energy level be filled with unlimited numbers of particles. The implications for these different statistics— for which there is no classical analogon—has been hinted at in Sect. 8.6.1.2 and is discussed extensively in books on statistical mechanics, for example, Ter Haar (1966).

The measurement of the spins on a two-particle system that is prepared to be in a 'singlet' state, that is, the system has zero intrinsic angular momentum is then such that when the spin on one of the particles is measured in the z-direction, then a measurement of the spin of the other particle gives exactly the opposite. This holds no matter how far apart the particles are. This is what has made physicists so worried about it: it appears that there are *non-local effects* in quantum mechanics, which, moreover, 'work' instantaneously, unencumbered by considerations of the speed of light. If the experiment is repeated many times the number of ups and downs measured on the first particle may be either; there is no way of telling which one it

will be. Therefore, this system cannot be used to 'send' a definite signal from one
particle to the other. There is cause for uneasiness though, as Einstein expressed, at
these non-local effects.

Instead of measurements on electrons, photonic experiments can be done. The
polarisation of light is easily measured using polaroid plates. An exact analogon
holds between the electron spin and the light polarisation; to make it work the light
has to be circularly polarised. For details of this phenomenon the reader is referred
to the literature, for example, Goswami (1992) or Sakurai (1985). The analysis of
these experiments has then guided physicists to exactly how quantum mechanics is
different from classical mechanics.

8.9.6 Quantum Mechanics as a Non-counterfactual-definite Theory, Bell's Inequality

There are then two issues that require elucidation when it comes to the interpretation
of quantum mechanics and its relation to measurement. The first is that the state of a
system is not determined until it is measured; this is essentially the Schrödinger's cat
problem. The second pertains to superluminal communication. Until the late 1970s,
these two issues were largely swept under the carpet, despite Einstein, Podolsky and
Rosen's paper in 1935, Einstein et al. (1935), in which these physically counter-
intuitive features of quantum mechanics were first flagged up. Ultimately, in physics
experiment is the arbiter. An idea by John Bell has delivered the relevant answer,
though it is a somewhat unusual way ahead.

In order to deliver on this issue the exact role of measurement in quantum mechanics needs to be understood. The key question is whether the information that is stored
in the wave function is *all that is available* or if there is information in the system
that is in some sense concealed. In other words, are there *pre-existing properties* in
the system, or—to use an anthropomorphism—does the system itself have no idea
what state it is in. A so-called *counterfactual-definite theory* is a theory in which
measurement reveals pre-existing properties. Classical mechanics is counterfactual-
definite: objects do not acquire properties which were, in some sense, ambiguous
before the measurement took place. John Bell discovered that it is possible to say
something about a series of measurements, assuming that the objects are classical,
that can be phrased in terms of an *inequality*. A series of experiments may then be
carried out to see if this inequality holds for measurements on a quantum system. If
not, then the classical assumption of counterfactual-definiteness does not hold and
the wave function contains all information without having to take recourse to hidden
variables. If, on the other hand, the inequality *does* hold, then Einstein's assumption
that there must be hidden variables is true.

8.9.6.1 Bell's Inequality

One version of Bell's inequality, published in 1964 (Bell 1964), has been presented by Maccone (2013); in this paper, a quantum system in which the inequality is violated is also presented. This is the treatment that will be followed here. Other treatments are available, for instance, a very lucid one by Bernard d'Espagnat (1979).

Maccone considers classical objects that have three pre-determined properties (this ensures a counterfactual-definite result of the theory). These properties are called A, B and C and they may be either $+$ or $-$. It is easy to think of examples. Maccone thinks of coins, which may be either gold ($+$) or copper ($-$) as property A. Property B says whether the coin is dull ($+$) or shiny ($-$) and property C is concerned with its size: large ($+$) or small ($-$). Another example could be towels that can be large or small, smooth or fluffy and blue or white. Whatever example of a classical object is chosen, each object is characterised by the triad (A, B, C). For example, (A^+, B^-, C^+) would pertain to a large, fluffy, blue towel. What is important is that the objects have the properties intrinsically ('they bring them with them') and not— as quantum mechanics would suggest—as features that become apparent at the point of measurement, or in the action of measurement.

Bell's inequality is concerned with the probability of measuring two of the three properties the same. For example, if A and B are measured the probability of a combination of A^+, B^+ and A^-, B^- is established. This probability is called $P_{\text{same}}(A, B)$. The probability that they are not the same, that is the combinations A^+, B^- and A^-, B^+ is denoted by $P_{\text{diff}}(A, B)$ and obviously $P_{\text{diff}}(A, B) = 1 - P_{\text{same}}(A, B)$. Bell's inequality states that

$$P_{\text{same}}(A, B) + P_{\text{same}}(A, C) + P_{\text{same}}(B, C) \geq 1 \qquad (8.122)$$

Maccone uses a graphical method with Venn diagrams to prove this for classical objects (which obey counterfactual-definite theory). Here, a more algebraic method is pursued. If a tremendously long series of observations is carried out, the number of hits that give a combination of the three properties A, B and C having a certain $+$ or $-$ value is denoted by $N(A^\bullet, B^\bullet, C^\bullet)$, where $\bullet$ is either $+$ or $-$. If $\mathcal{N}$ measurements are carried out, the value of $P_{\text{same}}(A, B)$ is

$$P_{\text{same}}(A, B) = \frac{N(A^+, B^+, C^+) + N(A^-, B^-, C^+) + N(A^+, B^+, C^-) + N(A^-, B^-, C^-)}{\mathcal{N}} \qquad (8.123)$$

Similarly,

$$P_{\text{same}}(A, C) = \frac{N(A^+, B^+, C^+) + N(A^+, B^-, C^+) + N(A^-, B^+, C^-) + N(A^-, B^-, C^-)}{\mathcal{N}} \qquad (8.124)$$

For the combination $(B^\bullet, C^\bullet)$, the probability $P_{\text{diff}}(B, C)$ is evaluated

$$P_{\text{diff}}(B, C) = \frac{N(A^+, B^+, C^-) + N(A^+, B^-, C^+) + N(A^-, B^+, C^-) + N(A^-, B^-, C^+)}{\mathcal{N}} \qquad (8.125)$$

Now, using $P_{\text{diff}}(B, C) = 1 - P_{\text{same}}(B, C)$, the quantity $P_{\text{same}}(A, B) +$
$P_{\text{same}}(A, C) + P_{\text{same}}(B, C)$ may be evaluated

$$P_{\text{same}}(A, B) + P_{\text{same}}(A, C) + P_{\text{same}}(B, C) = 1 +$$

$$+ \frac{N(A^+, B^+, C^+) + N(A^-, B^-, C^+) + N(A^+, B^+, C^-) + N(A^-, B^-, C^-)}{\mathcal{N}}$$

$$+ \frac{N(A^+, B^+, C^+) + N(A^+, B^-, C^+) + N(A^-, B^+, C^-) + N(A^-, B^-, C^-)}{\mathcal{N}}$$

$$- \frac{N(A^+, B^+, C^-) + N(A^+, B^-, C^+) + N(A^-, B^+, C^-) + N(A^-, B^-, C^+)}{\mathcal{N}}$$

$$= 1 + 2 \frac{N(A^+, B^+, C^+) + N(A^-, B^-, C^-)}{\mathcal{N}} \tag{8.126}$$

Neither $N(A^+, B^+, C^+)$ nor $N(A^-, B^-, C^-)$ can ever be negative and therefore
(8.122) follows.

Maccone then goes on to give an example of a quantum system that *violates* the
inequality. In this example, the three properties of the quantum system are characterised by eigen vectors. He chooses

$$A : \rightarrow \begin{bmatrix} |a_0\rangle = |0\rangle \\ |a_1\rangle = |1\rangle \end{bmatrix} \quad B : \rightarrow \begin{bmatrix} |b_0\rangle = \frac{1}{2}|0\rangle + \frac{\sqrt{3}}{2}|1\rangle \\ |b_1\rangle = \frac{\sqrt{3}}{2}|0\rangle - \frac{1}{2}|1\rangle \end{bmatrix} \quad C : \rightarrow \begin{bmatrix} |c_0\rangle = \frac{1}{2}|0\rangle - \frac{\sqrt{3}}{2}|1\rangle \\ |c_1\rangle = \frac{\sqrt{3}}{2}|0\rangle + \frac{1}{2}|1\rangle \end{bmatrix}$$

$$\tag{8.127}$$

The states within each pair are orthogonal. Note how B and C can be obtained from A
by a rotation in ket space. Now construct the two-level systems in the joint entangled
state

$$|\Phi^+\rangle = \frac{|00\rangle + |11\rangle}{\sqrt{2}} \tag{8.128}$$

(the $\sqrt{2}$ is necessary to normalise the total probability to unity). This should enable
the evaluation of the probabilities $P_{\text{same}}(\bullet, \bullet)$.

In order to evaluate $|00\rangle$, write $|0\rangle = \langle b_0|0\rangle |b_0\rangle + \langle b_1|0\rangle |b_1\rangle$. The projections
$\langle b_0|0\rangle$ and $\langle b_1|0\rangle$ can be read from (8.127); they are $\langle b_0|0\rangle = 1/2$ and $\langle b_1|0\rangle = \sqrt{3}/2$.
In a similar way, $|11\rangle$ is expanded in terms of $|b_0\rangle$ and $|b_1\rangle$. Altogether $|\Phi^+\rangle$ takes
the form

$$|\Phi^+\rangle = \frac{|a_0\rangle \left(|b_0\rangle + \sqrt{3}|b_1\rangle \right) + |a_1\rangle \left(\sqrt{3}|b_0\rangle - |b_1\rangle \right)}{2\sqrt{2}} \tag{8.129}$$

Selecting the $|a_0\rangle|b_0\rangle$ and $|a_1\rangle|b_1\rangle$ terms absolute squared yields $P_{\text{same}}(A, B)$. It
follows that $P_{\text{same}}(A, B) = 1/8 + 1/8 = 1/4$.

For $P_{\text{same}}(A, C)$, the following form of $|\Phi^+\rangle$ is relevant (note the projections
$\langle c_0|0\rangle = 1/2$ and $\langle c_1|0\rangle = -\sqrt{3}/2$)

$$|\Phi^+\rangle = \frac{|a_0\rangle \left(|c_0\rangle + \sqrt{3}|c_1\rangle\right) - |a_1\rangle \left(\sqrt{3}|c_0\rangle - |c_1\rangle\right)}{2\sqrt{2}} \tag{8.130}$$

And selecting the terms $|a_0\rangle|c_0\rangle$ and $|a_1\rangle|c_1\rangle$, again absolute squared, yields $P_{same}(A, C)$. It follows that $P_{same}(A, C) = 1/8 + 1/8 = 1/4$.

The form of $|\Phi^+\rangle$ relevant to the combination B—C is

$$|\Phi^+\rangle = \frac{\left(|b_0\rangle + \sqrt{3}|b_1\rangle\right)\left(|c_0\rangle + \sqrt{3}|c_1\rangle\right) - \left(\sqrt{3}|b_0\rangle - |b_1\rangle\right)\left(\sqrt{3}|c_0\rangle - |c_1\rangle\right)}{4\sqrt{2}}$$

$$\tag{8.131}$$

From this it follows that $P_{same}(B, C) = 1/4$. Altogether then the terms in Bell's inequality are

$$P_{same}(A, B) + P_{same}(A, C) + P_{same}(B, C) = \frac{3}{4} \tag{8.132}$$

which is clearly less than unity, so the inequality is violated.

The experimental test of Bell's inequality has presented a few problems. Nevertheless, by using the equivalence of spin states and circularly polarised light a successful series of experiments, initially done by Clauser and Freedman (1972), and refined by Aspect (1982), d'Espagnat (1979), and Zeilinger (1998), (three of whom won the Nobel prize for this work in 2022). It was found that quantum particles are *not* classical objects, as they violate Bell's inequality. Therefore, quantum mechanics involves an element of non-locality. Exactly how this aspect should be incorporated into the theory is not established in such a way that it can be said that there is a consensus. Despite this, the whole—some would say, audacious—development of the ideas demonstrates yet another distinct deviation from classical mechanics. At the same time, it shows that the wave function is the only information that is available. There are no hidden variables.

There is something intriguing about Bell's inequality. It is clearly based on a notion that classical objects come with pre-defined properties, like the examples of the coins or the towels used above. However, it is *not* based on classical *mechanics*. There is no limit that says: 'if this or that limit is taken, Newtonian mechanics follows'; there is no motion. The notion is one of a classical idea of information, or properties attached to objects. So, quantum mechanics has moved from a theory of motion (describing particles as material waves with momentum and position intertwined) to a theory of information. It could be argued that this was already clear when the probability interpretation was adopted. However, Bell's ideas put this aspect of the theory in a much brighter light.

References

Abramowitz M, Stegun A (1972) Handbook of mathematical functions. Dover, New York

Aspect A, Dalibard J, Roger G (1982) Experimental test of Bell's inequalities using time-varying analyzers. Phys Rev Lett 49(25): 1804–1807. Bibcode:1982PhRvL..49.1804A. https://doi.org/10.1103/physrevlett.49.1804

Bell JS (1964) On the Einstein Podolsky Rosen paradox. Physics Physique Fizika 1(3):195–200. https://doi.org/10.1103/PhysicsPhysiqueFizika.1.195

Bes DR (2013) Quantum mechanics. Springer, Berlin. https://doi.org/10.1007/978-3-662-05384-3.

Boas ML (1983) Mathematical methods in the physical sciences. Wiley, New York

Einstein A, Podolsky B, Rosen N (1935) Can quantum-mechanical description of physical reality be considered complete? Phys Rev 47(10):777–780

d'Espagnat B (1979) The quantum theory and reality. Sci Am 241(5):158–181

Farmelo G (2009) The strangest man: the hidden life of Paul Dirac, quantum genius. Faber and Faber, London

Freedman SJ, Clauser JF (1972) Experimental test of local hidden-variable theories. Phys Rev Lett 28(938):938–941. https://doi.org/10.1103/PhysRevLett.28.93

Gasiorowicz S (1995) Quantum physics, 2nd edn. Wiley, New York

Goswami A (1992) Quantum mechanics. Wm. C. Brown, Dubuque

Hofmann A (2018) Physical chemistry essentials. Springer, Cham

Jackson JD (1962) Classical elctrodynamics. Wiley, New York

Landau LD, Lifshitz EM (1976) Quantum mechanics, Course of theoretical physics, vol 3. Pergamon, Oxford

Maccone L (2013) A simple proof of Bell's inequality. Am J Phys 81:854–859. https://doi.org/10.1119/1.4823600

Mandl F (1988) Statistical physics, 2nd edn. Wiley, Chichester

Merzbacher E (1998) Quantum mechanics, 3rd edn. Wiley, New York

McCormmach R (1982) Night thoughts of a classical physicist. Harvard University Press, Cambridge

McMurry SM (1994) Quantum mechanics. Addison-Wesley, Wokingham

Moelwyn-Hughes EA (1966) A short course on physical chemistry. Longman, London

Pan J-W, Bouwmeester D, Weinfurter H, Zeilinger A (1998) Experimental entanglement swapping: entangling photons that never interacted. Phys Rev Lett 80:3891–3894. https://doi.org/10.1103/PhysRevLett.80.3891

Sakurai JJ (1985) Modern quantum mechanics. Benjamin/Cummings, Menlo Park

Schmitt A (2014) Introduction to superfluidity. Springer, Cham

Ter Haar D (1966) Elements of thermostatistics, 2nd edn. Holt, Rinehart and Winston, New York

Van Kampen NG (1988) Then theorems about quantum mechanical measurements. Physica A 153:97–113

Appendix A
Mathematical Appendix

This appendix contains material that will largely be known to the undergraduate maths user. It has been included here mostly as an aide-memoire. It also touches on aspects that are not generally well known, such as Bessel functions. A general text on the relevant elements used in this book is Boas (1983). Very useful furthermore is Abramowitz and Stegun (1972). For historical interest, Descartes (2009) is included in the bibliography.

A.1 Cartesian Vectors

Cartesian vectors are denoted by a bold facetype. They have as many components as the dimension of the space in which they operate. In three dimensions, the vector $\mathbf{x}$ has components x_1, x_2, x_3 (frequently denoted by x, y, z). It transforms under a rotation $\mathbf{Q}$ as $x_i' = Q_{ij} x_j$, where Einstein's summation convention has been used: twice occurring subscripts are summed over. The rotation has the properties that the determinant $\det \mathbf{Q} = 1$ and its inverse is its transposed: $Q^{-1} = Q^T$, where $Q_{ij}^{-1} Q_{jk} = \delta_{ik}$; δ_{ik} is the Kronecker delta (the identity): $\delta_{ij} = 1$ when $i = j$ and $\delta_{ij} = 0$ when $i \neq j$.

The *inner product* of two Cartesian vectors $\mathbf{a}$ and $\mathbf{b}$ is $a_i b_i$. It is invariant under rotation: $r_i' p_i' = Q_{ij} r_j Q_{ik} p_k = Q_{ji}^T Q i k r_j p_k = Q_{ji}^{-1} Q i k r_j p_k = \delta_{jk} r_j p_k = r_j p_j$. Another notation for the inner product is $\mathbf{a} \cdot \mathbf{b} = a_i b_i$.

The inner product is equal to the product of the magnitude of the vectors times the cosine of the angle between them.

The magnitude of a vector $\mathbf{r}$ is $\sqrt{r_i r_i}$.

The *Levi-Civita tensor* is ϵ_{ijk}, where $\epsilon_{123} = 1$ and any even permutation of the subscripts multiplies it by -1 and when any two subscripts are equal the outcome is 0.

The Levi-Civita tensor is used to define an outer product of two vectors $\mathbf{a}$ and $\mathbf{b}$: $c_i = \epsilon_{ijk} a_j b_k$. The outcome is another vector $\mathbf{c}$. The outer product is a vector with

M. A. C. Koenders, *Constructing the Edifice of Mechanics*, Undergraduate Texts in Physics, https://doi.org/10.1007/978-3-031-34071-0

Fig. A.1 Left: summing two vectors by laying one after the other. Right: illustration of the outer product

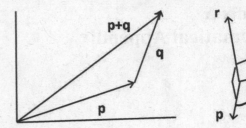

a magnitude that is a product of the magnitudes of the two vectors times the sine of the angle between them and it points in the direction that is normal to both the participating vectors and has the direction that fits like a 'right-handed screw' when going from **a** to **b**; Fig. A.1-right gives an illustration.

An alternative notation for the outer product of two vectors **a** and **b** is $\mathbf{c} = \mathbf{a} \times \mathbf{b}$. A theorem concerning the outer product is that the double outer product has the form

$$\mathbf{a} \times (\mathbf{b} \times \mathbf{c}) = \mathbf{b}(\mathbf{a}.\mathbf{c}) - \mathbf{c}(\mathbf{a}.\mathbf{b}) \tag{A.1}$$

This rule is sometimes called the 'bac-cab product'.

Vectors can be conveniently represented as arrows. Summing two vectors is then equivalent to laying one behind the other, see Fig. A.1-left. The components are simply the sums of the participating vector components: $r_i = p_i + q_i$.

In four dimensions (so-called *four-vectors*), no bold-facetype is applied and the transformation is the Lorentz transformation L, see Sect. 6.3, which includes the three-dimensional rotations. The components are denoted by Greek subscripts, so $x'_\mu = L_{\mu\nu}x_\nu$. The determinant of L is again equal to 1 and its inverse is its transposed.

A.2 Ket Vectors

A Hilbert space is a system of elements (also called 'states') $|\psi\rangle$, $|\phi\rangle$, $|\chi\rangle$, ... with the following properties:

The Hilbert space is a linear space with rules such as:

$|\psi\rangle + |\phi\rangle = |\phi\rangle + |\psi\rangle$;

$\lambda[|\psi\rangle + |\phi\rangle] = \lambda|\psi\rangle + \lambda|\phi\rangle$;

$(\lambda + \mu)|\psi\rangle = \lambda|\psi\rangle + \mu|\psi\rangle$, etc.

Note that numbers are complex.

For every pair $|\psi\rangle$ and $|\phi\rangle$, there is an inner product $\langle\phi|\psi\rangle$, with the following properties:

if $|\chi\rangle = \lambda|\phi\rangle$ then:

$\langle\chi|\psi\rangle = \lambda^*\langle\phi|\psi\rangle$;

$\langle\psi|\chi\rangle = \langle\chi|\psi\rangle^* = \lambda\langle\psi|\phi\rangle$;

if $|\phi\rangle = |\phi_1\rangle + |\phi_2\rangle$, then $\langle\phi|\psi\rangle = \langle\phi_1|\psi\rangle + \langle\phi_2|\psi\rangle$ and

$\langle\phi|\phi\rangle \geq 0$ (the equal sign holds when $|\phi\rangle = 0$).
The norm $||\phi||$ is defined as $||\phi|| = \sqrt{\langle\phi|\phi\rangle}$.
Other theorems relevant to ket vectors are discussed in Sect. 8.7.1.

A.3   Differential Field Operators

The *gradient* of a function $f(x, y, z)$ is defined as a vector $\mathbf{V}$ of the form

$$(V_1, V_2, V_3) = \text{grad } f(x, y, z) = \left(\frac{\partial f}{\partial x}, \frac{\partial f}{\partial y}, \frac{\partial f}{\partial z}\right) \tag{A.2}$$

A frequently used notation makes use of the so-called 'nabla'-vector, defined as the vector operator

$$\nabla = (\frac{\partial}{\partial x}, \frac{\partial}{\partial y}, \frac{\partial}{\partial z}) \tag{A.3}$$

In this notation, the gradient takes the form grad $f = \nabla f$. It is also possible to express the gradient in components as $V_i = \partial f/\partial x_i$.

Note that $f(x, y, z)$ is a scalar function and ∇f is a vector.

The *divergence* of a vector function field $f_1(x, y, z)$, $f_2(x, y, z)$, $f_3(x, y, z)$ is defined as

$$\text{div } \mathbf{f} = \sum_{i=1}^{3} \frac{\partial f_i}{\partial x_i} \tag{A.4}$$

Alternative notations are div $\mathbf{f} = \nabla \cdot \mathbf{f}$ or $\partial f_i/\partial x_i$ (use Einstein's summation convention).

The divergence operates on a vector function and its result is a scalar.

The *curl* (or the rotation) of a vector field $\mathbf{f}(x, y, z)$ is an outer product

$$\text{curl } \mathbf{f} = \nabla \times \mathbf{f}; \text{ the } i\text{th component is } \epsilon_{ijk}\frac{\partial f_j}{\partial x_k} \tag{A.5}$$

The curl operator works on a vector field and the result is another vector field.

There are various theorems that are quite easy to prove

$$\text{curl grad } f(x, y, z) = 0 \tag{A.6}$$

$$\text{div curl } \mathbf{f}(x, y, z) = 0 \tag{A.7}$$

Furthermore, the *Laplacian* operator is defined as

$$\Delta f(x, y, z) = \text{div grad } f(x, y, z) = \nabla^2 f(x, y, z) = \frac{\partial^2 f(x, y, z)}{\partial x_i^2} \tag{A.8}$$

Using the bac-cab product, it is shown that

$$\triangle f = \text{grad div} f - \text{curl curl } f \tag{A.9}$$

The Laplacian can also be defined for four-vectors, in which case the definition is (using the summation convention)

$$\Box f(x_1, x_2, x_3, x_4) = \frac{\partial^2 f}{\partial x_\mu^2} \tag{A.10}$$

Instead of the notation $\Box$, the notation $\nabla^{(4)}$ is sometimes employed.

A.3.1 Theorems Involving Integrals and Differential Operators

Theorems involving integrals and differential field operators are as follows.

Gauss' theorem (sometimes called the integral divergence theorem), which concerns an integral over a volume V, which has a surface S; an outward unit normal to the surface is called $\mathbf{n}$

$$\iiint_V \text{div } \mathbf{f} dV = \oiint_S \mathbf{f} \cdot \mathbf{n} dS \tag{A.11}$$

Similarly, *Stokes' theorem* is concerned with an integral over a closed loop L, which has a surface S. Going round the loop in vector increments, $d\mathbf{L}$ defines a right-handed screw orientation; the unit normal on the surface spanned by the loop $\mathbf{n}$ is its direction.

$$\iint_S \text{curl } \mathbf{f} \cdot \mathbf{n} dS = \oint_L \mathbf{f} \cdot d\mathbf{L} \tag{A.12}$$

A.4 Lagrange Multipliers

A function of the variables $q_1...q_n$ is $f(q_1, q_2, q_3...q_n)$. The question is: when is this function extremal (that is, has a maximum or a minimum) on the condition that a certain constraint is satisfied? The constraint may be phrased as a function of the variables $q_1...q_n$ as $G(q_1, q_2, q_3...q_n) = 0$. Lagrange introduced a multiplier λ and constructed the functional

$$F(q_1, q_2, q_3...q_n) = f(q_1, q_2, q_3...q_n) + \lambda G(q_1, q_2, q_3...q_n) \tag{A.13}$$

and required

$$\frac{\partial F}{\partial q_i} = 0; \quad i = 1...n \tag{A.14}$$

A simple example best illustrates how this works. The task is to make a rectangular box of volume V with sides a, b, c that minimises the area A of sheet material from which the box is made. The constraint is now $abc - V = 0$; the function to be 'extremalised' is $f(a, bc) = 2ab + 2ac + 2bc$ (the area of the box). Now construct the functional

$$F(a, b, c) = 2ab + 2ac + 2bc + \lambda(abc - V) \tag{A.15}$$

Differentiating gives

$$\frac{\partial F}{\partial a} = 2b + 2c + \lambda bc = 0; \quad \frac{\partial F}{\partial b} = 2a + 2c + \lambda ac = 0; \quad \frac{\partial F}{\partial c} = 2a + 2b + \lambda ab = 0 \tag{A.16}$$

It is seen immediately that the solution is $a = b = c$ and that $a = -4/\lambda$. The solution is a cube as the rectangular box that minimises surface area for a given volume. The constraint may now be used to determine $\lambda = -4V^{-1/3}$ and it follows that $a = V^{1/3}$. Even though this example is trivial it shows how the method works.

There need not be merely one constraint. The method works just as well for a number of constraints $G_j(q_1, q_2, q_3...q_n)$; obviously, there need to be fewer constraints than the dimension of the problem. For each constraint, a Lagrange multiplier is introduced, so the functional becomes

$$F = f(q_1, q_2, q_3...q_n) + \lambda_1 G_1(q_1, q_2, q_3...q_n) + \lambda_2 G_2(q_1, q_2, q_3...q_n) + ...\text{etc} \tag{A.17}$$

Then require as before

$$\frac{\partial F}{\partial q_i} = 0; \quad i = 1...n \tag{A.18}$$

A.5 Fourier Series

A periodic function $f(t)$ with period T (frequency $f = 1/T$; circular frequency $\omega = 2\pi/T$) can be written as an infinite sum of sines and cosines, as follows:

$$f(t) = a_0 + \sum_{n=1}^{\infty} \left[a_n \cos\left(\frac{2n\pi t}{T}\right) + b_n \sin\left(\frac{2n\pi t}{T}\right) \right] \tag{A.19}$$

Here, a_0, a_n and b_n are the *Fourier coefficients*. They may be derived from the function $f(t)$ as

$$a_0 = \frac{1}{T} \int_{-T/2}^{T/2} f(t) dt \tag{A.20}$$

$$a_n = \frac{1}{T} \int_{-T/2}^{T/2} \cos\left(\frac{2\pi t}{T}\right) f(t)dt; \quad b_n = \frac{1}{T} \int_{-T/2}^{T/2} \sin\left(\frac{2\pi t}{T}\right) f(t)dt \qquad \text{(A.21)}$$

a_0 is just the average of the function $f(t)$. a_n and b_n follow the periodic part and, as their frequencies become higher, they follow the function more and more accurately.

A.6   Fourier Transformation and Delta Function

The Fourier transformed of a function $f(t)$ is $\hat{f}(\omega)$, defined as

$$\hat{f}(\omega) = \int_{-\infty}^{\infty} f(t)e^{-i\omega t}dt \qquad \text{(A.22)}$$

Its inverse is

$$f(t) = \frac{1}{2\pi} \int_{-\infty}^{\infty} \hat{f}(\omega)e^{i\omega t}d\omega \qquad \text{(A.23)}$$

The delta function is

$$\delta(\lambda) = \frac{1}{2\pi} \int_{-\infty}^{\infty} e^{i\tau\lambda}d\tau \qquad \text{(A.24)}$$

This is a form that does not make much sense unless it is used in an integral with a 'decent' function $g(t)$:

$$g(t) = \int_{-\infty}^{\infty} \delta(t - \tau)g(\tau)d\tau \qquad \text{(A.25)}$$

A.7   Laplace Transform

The Laplace transform is very similar to the Fourier transform. A function $f(t)$ has a Laplace transform $\hat{f}(s)$, denoted as $\text{Lap}[f(t)]$

$$\hat{f}(s) = \int_{0}^{\infty} f(t)e^{-st}dt \qquad \text{(A.26)}$$

The Laplace transform is generally very useful in solving differential equations. The back transform is not so easy, however, most practitioners use tables to identify the form. A good table is found in Abramowitz and Stegun (1972).

A.8 The Gamma Function, Error Function and Bessel Functions

A.8.1 Gamma Function

The gamma function $\Gamma(x)$ is defined as

$$\Gamma(x) = \int_0^\infty t^{x-1} e^{-t} dt \qquad (A.27)$$

If x is a natural number n, then

$$\Gamma(n+1) = n! \qquad (A.28)$$

Asymptotically for $n \to \infty$, Stirling's formula holds

$$n! \approx n^n e^{-n} \sqrt{2\pi n} \qquad (A.29)$$

The logarithm of the gamma function is depicted in Fig. A.2.

Fig. A.2 $\log\Gamma(x)$

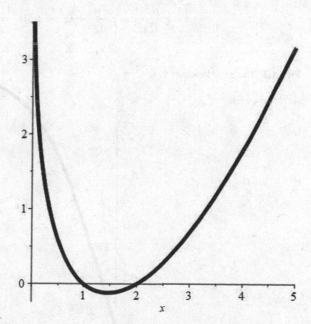

A.8.2 Error Function

The error function is defined as

$$\mathrm{erf}(x) = \frac{2}{\sqrt{2}} \int_0^x e^{-t^2} dt \qquad (A.30)$$

The complementary error function is $\mathrm{erfc}(x) = 1 - \mathrm{erf}(x)$. A useful Laplace transform involving the error function is

$$\mathrm{Lap}\left[\mathrm{erfc}\left(\frac{k}{2\sqrt{t}}\right)\right] = \frac{1}{s}e^{-k\sqrt{s}} \qquad (A.31)$$

An impression of the error function is provided in Fig. A.3.

A.8.3 Bessel Functions

Bessel functions are solutions of the differential equation

$$z^2\frac{d^2w}{dx^2} + z\frac{dw}{dz} + (z^2 - \nu^2)w = 0 \qquad (A.32)$$

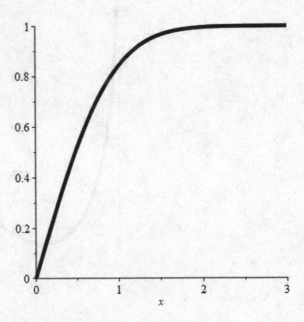

Fig. A.3 The error function

The solution is a special function $J_\nu(z)$ or $Y_\nu(z)$, where ν is the order of the Bessel function. $J_\nu(z)$ has the series expansion

$$J_\nu(z) = \left(\frac{z}{2}\right)^\nu \sum_{k=0}^{\infty} \frac{\left(-\frac{z^2}{4}\right)^k}{k!\,\Gamma(\nu+k+1)} \tag{A.33}$$

The zeroth- and first-order Bessel functions are depicted in Figs. A.4 and A.5. The functions $Y_\nu(z)$ also have a series expansion, but it is a rather complicated expression, see Abramowitz and Stegun (1972), where many more useful relations are listed. The asymptotic expansion for the Bessel functions are

$$J_\nu(z) \to \sqrt{\frac{2}{\pi z}} \cos\left(z - \frac{\nu z}{2} - \frac{\pi}{4}\right) + O\left(z^{-1}\right), \; z \to \infty \tag{A.34}$$

Modified Bessel functions satisfy the differential equation

$$z^2 \frac{d^2 w}{dx^2} + z\frac{dw}{dz} - (z^2 + \nu^2)w = 0 \tag{A.35}$$

The solutions are called $I_{\pm\nu}$ and $K_\nu(z)$. The series expansion for $I_\nu(z)$ is

Fig. A.4 Bessel functions. Solid line: $J_0(x)$; dashed line: $J_1(x)$. Note the oscillatory character for large argument

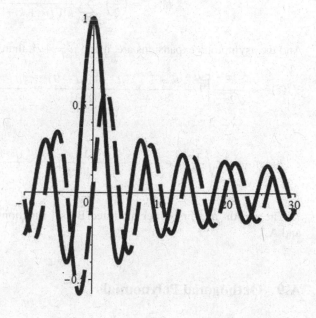

Fig. A.5 Bessel functions.
Solid line: $Y_0(x)$; dashed
line: $Y_1(x)$. Note the
oscillatory character for
large argument

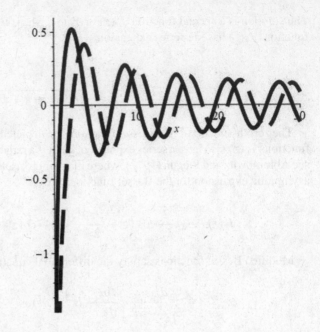

$$I_\nu(z) = \left(\frac{z}{2}\right)^\nu \sum_{k=0}^\infty \frac{\left(\frac{z^2}{4}\right)^k}{k!\,\Gamma(\nu + k + 1)} \tag{A.36}$$

And the asymptotic expansions are, using $\mu = 4\nu^2$, then for $z \to \infty$

$$I_\nu(z) \to \frac{e^z}{\sqrt{2\pi z}} \left[1 - \frac{\mu - 1}{8z} + \frac{(\mu - 1)(\mu - 9)}{2!(8z)^2} - \frac{(\mu - 1)(\mu - 9)(\mu - 25)}{3!(8z)^3} + ... \right] \tag{A.37}$$

$$K_\nu(z) \to e^{-z}\sqrt{\frac{\pi}{2z}} \left[1 + \frac{\mu - 1}{8z} + \frac{(\mu - 1)(\mu - 9)}{2!(8z)^2} + \frac{(\mu - 1)(\mu - 9)(\mu - 25)}{3!(8z)^3} + ... \right] \tag{A.38}$$

The zeroth- and first-order modified Bessel functions are depicted in Figs. A.6 and A.7.

A.9 Orthogonal Polynomials

These polynomials below have been studied very extensively and many other relations are available, see Boas (1983).

Fig. A.6 Modified Bessel functions. Solid line: $I_0(x)$; dashed line: $I_1(x)$. Note the exponential character for large argument

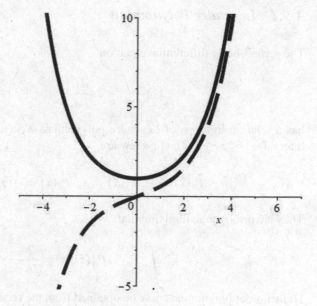

Fig. A.7 Modified Bessel functions. Solid line: $K_0(x)$; dashed line: $K_1(x)$. Note the $x^{-1/2}$ character for small argument

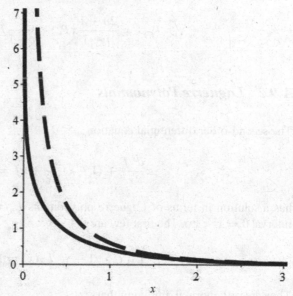

A.9.1 Legendre Polynomials

The second-order differential equation

$$(1 - x^2)\frac{d^2 f}{dx^2} - 2x\frac{df}{dx} + n(n+1)f = 0 \qquad (A.39)$$

has a solution in terms of Legendre polynomials $P_n(x)$. These are defined on the interval $-1 \le x \le 1$. The first few are

$$P_0(x) = 1; \quad P_1(x) = x; \quad P_2(x) = \frac{1}{2}(3x^2 - 1) \qquad (A.40)$$

They are orthogonal, implying that

$$\int_1^1 P_n(x) P_m(x) dx = \frac{2\delta_{mn}}{2n+1} \qquad (A.41)$$

Higher order polynomials may be obtained from the recursion relation

$$P_{n+1} = \frac{2n+1}{n+1} x P_n - \frac{n}{n+1} P_{n-1} \qquad (A.42)$$

A.9.2 Laguerre Polynomials

The second-order differential equation

$$x\frac{d^2 f}{dx^2} + (1 - x)\frac{df}{dx} + nf = 0 \qquad (A.43)$$

has a solution in terms of Laguerre polynomials $L_n(x)$. These are defined on the interval $0 = x < \infty$. The first few are

$$L_0(x) = 1; \quad L_1(x) = 1 - x; \quad L_2(x) = 1 - 2x + \frac{1}{2}x^2 \qquad (A.44)$$

They are orthonormal, implying that

$$\int_0^\infty e^{-x} L_n(x) L_m(x) dx = \delta_{mn} \qquad (A.45)$$

Higher order polynomials may be obtained from the recursion relation

$$L_{n+1} = \frac{2n+1-x}{n+1}L_n - \frac{n}{n+1}L_{n-1} \qquad (A.46)$$

A.9.3 Hermite Polynomials

The second-order differential equation

$$\frac{d^2 f}{dx^2} - 2x\frac{df}{dx} + 2nf = 0 \qquad (A.47)$$

has a solution in terms of Hermite polynomials $\mathcal{H}_n(x)$. These are defined on the interval $-\infty < x < \infty$. The first few are

$$\mathcal{H}_0(x) = 1; \quad \mathcal{H}_1(x) = 2x; \quad \mathcal{H}_2(x) = 4x^2 - 2 \qquad (A.48)$$

They are orthogonal, implying that

$$\int_{-\infty}^{\infty} e^{-x^2} \mathcal{H}_n(x)\mathcal{H}_m(x)dx = \sqrt{2}2^n n!\delta_{mn} \qquad (A.49)$$

Higher order polynomials may be obtained from the recursion relation

$$\mathcal{H}_{n+1} = 2x\mathcal{H}_n - 2n\mathcal{H}_{n-1} \qquad (A.50)$$

References

Abramowitz M, Stegun A (1972) Handbook of mathematical functions. Dover, New York
Boas ML (1983) Mathematical methods in the physical sciences. Wiley, New York
Descartes R (2009) Discours de la méthode. In: Œuvres complètes, Vol III: Discours de la Mèthode et Essais; eds. Jean-Marie Beyssade and Denis Kambouchner. Gallimard, Paris

Index

M. A. C. Koenders, *Constructing the Edifice of Mechanics*, Undergraduate Texts in Physics, https://doi.org/10.1007/978-3-031-34071-0

Printed in the United States
by Baker & Taylor Publisher Services